ALEXANDER STROH

Reactive Control of Turbulent Wall-Bounded Flows for Skin Friction Drag Reduction

SCHRIFTENREIHE DES INSTITUTS FÜR STRÖMUNGSMECHANIK
KARLSRUHER INSTITUT FÜR TECHNOLOGIE (KIT)

BAND 4

Reactive Control of Turbulent Wall-Bounded Flows for Skin Friction Drag Reduction

BY
ALEXANDER STROH

KIT Scientific Publishing

Dissertation, Karlsruher Institut für Technologie (KIT)
KIT-Fakultät für Maschinenbau

Tag der mündlichen Prüfung: 11. Dezember 2015
Referenten: Prof. Dr.-Ing. B. Frohnapfel, Prof. Dr. P. Schlatter

Impressum

KIT Scientific Publishing

Karlsruher Institut für Technologie (KIT)
KIT Scientific Publishing
Straße am Forum 2
D-76131 Karlsruhe

KIT Scientific Publishing is a registered trademark
of Karlsruhe Institute of Technology.
Reprint using the book cover is not allowed.

www.ksp.kit.edu

Print on Demand 2018 – Gedruckt auf FSC-zertifiziertem Papier

ISSN 2199-8868
ISBN 978-3-7315-0766-6
DOI 10.5445/KSP/1000080529

Reactive Control of Turbulent Wall-Bounded Flows for Skin Friction Drag Reduction

Zur Erlangung des akademischen Grades
Doktor der Ingenieurwissenschaften
der Fakultät für Maschinenbau
Karlsruher Institut für Technologie (KIT)

genehmigte
Dissertation
von

Dipl.-Ing. Alexander Stroh

Tag der mündlichen Prüfung:	11. Dezember 2015
Hauptreferent:	Prof. Dr.-Ing. B. Frohnapfel
Korreferent:	Prof. Dr. P. Schlatter

Acknowledgements

First of all I would like to express my gratitude to my doctoral adviser, Prof. Dr.-Ing. Bettina Frohnapfel, whose expertise has made a considerable contribution to the content of this thesis. During my employment as a doctoral researcher, she contributed to a memorable research experience by providing me with the opportunity to develop my own individuality by being allowed to work with reasonable independence, supporting my attendance at various conferences and engaging me in new ideas. I was fortunate to have the opportunity to work with Prof. Dr. Yosuke Hasegawa (The University of Tokyo), who provided me with scientific guidance, fruitful discussions and valuable advise. I gained so much from his vast knowledge and scientific curiosity. I would like to thank Prof. Dr.-Ing. Bettina Frohnapfel and Prof. Dr. Yosuke Hasegawa for introducing me to the fascinating and challenging world of turbulence research, along with their outstanding mentorship.

A special word of thanks is owed to Prof. Dr. Philipp Schlatter (KTH Mechanics, Stockholm) for providing the numerical code and expert advice regarding the implementation of various control schemes. I am also particularly grateful to him for agreeing to be co-examiner for my PhD thesis and for giving me several opportunities to visit him in Stockholm to discuss my ongoing research.

Furthermore, I would like to thank Prof. Dr. Koji Fukagata (Keio University, Tokyo) for the opportunity to spend several months in Japan at his laboratory and supporting my research on the topic of suboptimal control.

I am truly grateful to my student M. Sc. Juan Ignacio Polanco, whose work made a significant contribution to my research.

I would also like to thank my colleagues Dr.-Ing. Gerti Daschiel, Dr.-Ing. Andreas Güttler, Dr.-Ing. Davide Gatti, Dipl.-Ing. Tobias Schumm, M. Sc. Verena Fink, Dipl.-Ing. Anna Slotosch, M. Sc. Giacomo Cocconi, M. Sc. Andrea Codrignani, Dr. Pourya Forooghi and M. Sc. Frieder Kaiser for making my time at ISTM a lot of fun, shaping the friendly and cooperative everyday atmosphere at work and also providing useful feedback and insightful comments on my work.

I gratefully acknowledge the constant support and love of my parents, Alexander and Lubov, which they have provided throughout my entire life. In conclusion, I wish to express how grateful I am to Luba for her love, understanding, patience and motivating support for my work over the last few challenging years.

Abstract

The present work is an attempt to improve the understanding of reactive drag reducing control of near-wall turbulence with respect to limitations present in a realistic application scenario. For this purpose, the reference reactive control scheme of opposition control is investigated using direct numerical simulation (DNS) through the consideration of various application oriented restrictions, such as local control application, limited spatial and temporal resolution of the scheme, sensor noise and control elements arrangement.

The impact of the limitations on the control performance is evaluated in the configuration of a fully developed turbulent channel flow (TCF). A series of parametric studies is carried out for the identification of limiting values for spatial and temporal properties of the control scheme, as well as the noise polluted sensor signal. The challenging issue of a more realistic arrangement and placement of sensors and actuators is examined using a correlation based approach, which allows the derivation of modified control schemes with improved performance.

Furthermore, the realistic limitations linked to the necessity of local control application are investigated and discussed in the more suitable framework of a spatially developing turbulent boundary layer (TBL). The differences between control application in TBL and TCF in terms of control mechanism are identified and analysed using a mathematical decomposition of the skin friction coefficient. Furthermore, the characteristic effects of control activation in TBL are examined by applying localised body force damping and compared with the well known drag reducing scheme of uniform blowing. The integral performance indices linked to the alteration of TBL downstream of the control region are evaluated and discussed in detail. Based on the results obtained, a simple model for the estimation of the global development of controlled TBL is proposed and compared with simulation results.

Kurzfassung

Im Rahmen der vorliegenden Arbeit wird ein Beitrag zum Verständnis über die reibungsmindernde reaktive Kontrolle der wandnahen Turbulenz in Bezug auf eine realitätsnahe Anwendung geleistet. Zu diesem Zweck werden folgende anwendungsorientierte Limitierungen in die direkte numerische Simulation (DNS) der Referenzkontrollmethode "Opposition Control" [18] implementiert und getestet: lokaler Einsatz der Kontrolle, zeitlich und räumlich beschränkte Auflösung der Kontrollmethode, Sensorrauschen und Anordnung von Kontrollelementen.

Der Einfluss der eingeführten Limitierungen auf die Leistung der Kontrolle wird in der Konfiguration einer vollentwickelten turbulenten Kanalströmung (TCF) untersucht. Eine Reihe von Parameterstudien wird zur Identifizierung der zeitlichen und räumlichen Grenzbereiche der Kontrolle sowie der Effekte des Sensorrauschens durchgeführt. Die Problematik einer realitätsnäheren räumlichen Verteilung von Sensoren und Aktuatoren wird mit Hilfe einer Korrelationsanalyse untersucht. Auf Grundlage der Ergebnisse werden die Kontrollschemata modifiziert, was eine gesteigerte Kontrollleistung ermöglicht.

Die Limitierungen bezüglich des lokalen Einsatzes der Kontrolle werden in geeigneter Konfiguration einer räumlich entwickelnden turbulenten Grenzschicht (TBL) untersucht. Unterschiede bezüglich Kontrollmechanismen bei der Anwendung in TBL und TCF werden anhand einer Analyse der Zerlegung des Reibungskoeffizienten identifiziert und erläutert. Weiterhin werden die grenzschicht-charakteristischen Effekte der Kontrolle durch Anwendung einer weiteren Kontrollmethode namens "Body Force Damping" [54] untersucht und mit den Effekten, die durch gleichmäßiges Einblasen von Fluid in die Grenzschicht entstehen, verglichen. Die Änderung der integralen Leistungsgrößen durch die Kontrollaktivierung in TBL werden ausgewertet und ausführlich diskutiert. Basierend auf den erzielten Ergebnissen, wird abschließend ein einfaches Modell zur Abschätzung der globalen Entwicklung der Grenzschicht im kontrollierten Zustand vorgeschlagen und anhand von Simulationsdaten validiert.

Contents

1 Introduction

1.1 Background

Modification of the natural flow behaviour is one of the fundamental research topics of fluid mechanics. A considerable amount of work has been put into investigations in this field, which has solved various engineering problems during the evolution of civil and military aviation in the 20th century. Originally, this topic was handled using analytic tools and experimental set-ups. Since the introduction of numerical simulations to the research community, the challenges of goal-oriented flow properties modification, or flow control, has experienced a reincarnation. Due to the new unique abilities presented by the numerical simulations, it has finally been possible to obtain a closer insight into the detailed physics of the flow. Eventually, evolving computational capacities enabled researchers to tackle the problem of turbulent flows, known for their complexity and unpredictability using direct numerical simulations (DNS).

The realities of the 21st century shifted the goals of flow control from the performance oriented engineering optimisation towards more efficient usage of the diminishing energy resources. Since the majority of engineering applications have to deal with wall-bounded turbulent flows, the reduction of turbulent skin friction drag has become of great interest ecologically and economically. This goal attracts special attention in the world-wide transport sector due to the potential for much more efficient energy utilisation in various thermo-fluids systems, such as airplanes, high-speed rail and motor vehicles, marine vessels and pipeline transportation of fluids. Various estimations reveal tremendous potential savings linked to the reduction of turbulent skin friction drag. Gad-el-Hak [40] estimates $1 billion annual fuel savings if 20% skin friction drag reduction could be realised on every aircraft world-wide. Reneaux [114] presents another estimation where a drag reduction of 1% on a large transport aircraft corresponds to 0.2% of its direct operating costs, which could mean ten additional passenger seats per flight. As for the marine vessels, Gollub *et al.* [43] mentions a possible annual saving of $10 billion if 10% drag reduction can be achieved in the ocean shipping industry.

Investigations over recent decades aimed at the reduction of skin friction drag using direct numerical simulations and experiments have introduced several new turbulence control techniques. Reactive flow control is one of the promising control methods that uses flow state information captured by *sensors* in order to estimate the drag reducing control input imposed by actuators. This flow control technique has been proven to be very efficient in different flow configurations and surface geometries. However, most investigations using reactive flow control were carried out using numerical simulations assuming idealistic models of sensors and actuators and therefore cannot directly represent a realistic application case.

1.2 Objectives and Procedure

Due to the significant technological progress and immense growth of the available computing power in the last decades of 20th century DNS has become one of the most important and widespread tools of turbulence research [96, 118]. Thereby the field of DNS application greatly expanded from the first simulations of isotropic turbulence in the beginning of 1970s [101] to the framework of wall-bounded turbulent flows in the late 1980s followed by more complex flow geometries later on. The first simulation of a fully developed turbulent channel flow (TCF) at low Reynolds number ($Re_\tau = 180$) has been conducted in 1987 by Kim *et al.* [75] and one year later the first study with a DNS of a spatially developing turbulent boundary layer (TBL) has been reported by Spalart [132]. Striving for more realistic flow configurations, turbulence researches managed to gradually increase Reynolds number of simulations up to Re_τ of the order of 10^3, which is considered to be in the range of Re_τ relevant for industrial applications [130]. The state of the art TCF simulations with $Re_\tau = 2000$ [49], $Re_\tau = 4000$ [87] and $Re_\tau \approx 5200$ [84] have been recently reported. An investigation of zero-pressure-gradient TBL in the Reynolds number range up to $Re_\theta \approx 6700$ corresponding to $Re_\tau = 2000$ has been recently published by Sillero, Jiménez and Moser [128].

The main feature of a DNS is a complete resolution of temporal and spatial scales in a turbulent flow field. This ability enables the reliable application of turbulence control schemes using DNS since the entire multi-scale physics of the manipulation process can be captured. However, in most cases, the nature of such flow field modification is based on fairly unrealistic assumptions directly linked to the properties of the DNS. Turbulence control schemes developed using DNS adopt high spatial and temporal resolution along with the unlimited availability of the flow field information throughout the simulation domain. Correspondingly, the schemes show theoretical possibilities of flow control rather than practically relevant application methods. In the present thesis an attempt is made to investigate various representative numerical reactive control schemes with respect to realistic limitations.

Application of active or reactive control schemes assumes the presence of sensors and actuators in the simulation domain. The dimensions of the hypothetical sensors and actuators in a DNS are only limited by the spatial resolution of the simulation and thus every grid node of the simulation domain can represent an independent sensor or actuator. The size of such elements is related to the smallest motions of a turbulent flow field and therefore is considered to be almost infinitely small. In spite of the continuous miniaturisation of available electronic hardware over recent decades, actual sensors and actuators have a certain finite size. Hence the influence of the sensor or actuator finite size on the control performance needs to be investigated. Similarly, limitations in terms of temporal properties of a DNS have to be considered. DNS resolves the smallest temporal scales of the turbulent motions and therefore many numerical control schemes utilise quasi-continuous sensing and actuation. This assumption is expected to have a definite impact on the control efficiency and therefore has to be taken into account.

The issue of sensor and actuator distribution in a simulation domain has to be addressed in detail. The majority of the numerical control schemes propose a placement of sensors or actuators that is not compatible with a realistic setup due to blockage or positioning issues. For instance, the scheme of opposition control [18] utilises wall-normal velocity information at a certain position close to the wall surfaces or schemes based on optimal control [1, 17] employ the entire velocity information of the flow domain in order to estimate control input. The possibility of control scheme improvement in terms of more realistic placement of control elements therefore has to be verified and tested in terms of control efficiency.

Sensing in terms of numerical simulation assumes a simple assignment of some particular value from one variable of the simulation to another. A real measurement is obviously a more complicated process that is always connected to the natural distortion of the acquired information and depends on the control technique and equipment. Therefore it is important to know the influence of polluted sensor signal on the resulting efficiency of a numerical control scheme.

Although from the engineering point of view it is more important to be able to modify outer developing turbulent flows, i.e., spatially developing turbulent boundary layers, reactive control of turbulent flows is mostly tested in a configuration of a fully developed turbulent channel flow. TCF remains a popular control configuration due to the fact that the near-wall turbulent dynamics and flow properties are assumed to be universal in wall-bounded flows. Additionally, a channel flow DNS is much more convenient in terms of computational resources because of the presence of streamwise periodicity in the simulation domain and hence the presence of an additional homogeneous direction leading to a much shorter simulation time. The streamwise spatial development of a TBL significantly increases the computational costs of such DNS, where much longer simulation domains have to be used in order to reach high Reynolds numbers. Nevertheless, the effect of the spatial development of the flow in TBL on the control efficiency has to be investigated in order to complete the knowledge base concerning turbulence control. Since in most engineering applications it is difficult to utilise control on the entire wall area, local control application in TBL will be tested for the investigation of spatial transients due to the control application and its influence on the efficiency. Based on this knowledge, suggestions about control placement in TBL with a fixed flat plate length will be made.

The present thesis attempts to establish a link between the numerical flow control schemes and their potential practical application. For this purpose, the limitations just described are implemented into the reactive control scheme of opposition control [18] using DNS of a TCF. A series of parametric studies with variation of the limiting temporal and spatial quantities is carried out. The control efficiency is evaluated and compared with the original numerical control scheme in order to contribute to the understanding of the certain limitation influence on the control process. Based on the results, the most important issues for the transfer of numerical control schemes to reality are identified. Several modifications of the scheme are proposed in order to take into account newly introduced limitations. Since an investigation of spatial transients is rather difficult in the configuration of TCF due to the inherent periodicity, the transient behaviour introduced by a locally applied control is explored in a spatially developing flow using DNS of a TBL. The control effects in TBL are compared with the effects in TCF in order to investigate the differences due to the streamwise development of the flow. Global effects of the local control application are studied using the body force damping scheme and then compared with the drag reducing scheme of uniform blowing. The differences in the flow behaviour downstream of the control region are identified and elucidated, followed by an introduction of a simplified model, which allows an estimation of the global control effect based on the control effect within the control region.

1.3 Outline

The thesis begins with an introduction of fluid mechanics backgrounds, a description of the turbulent wall-bounded flows, flow control methods and analytical tools utilised in the present investigation (Chapter 2).

The scheme of opposition control proposed by Choi *et al.* [18] is chosen as a representative numerical reactive control scheme due to its simplicity and high efficiency. Thus, this scheme is used as a reference for the modified control approaches presented throughout the thesis. Chapter 3 addresses the discrepancy between the conditions of the numerical simulation and realistic conditions in the configuration of a TCF. The chapter begins with a description of the DNS performed and a discussion of the difficulties linked to the limited domain size used in DNS. The rest of the chapter is dedicated to various aspects of the reactive control scheme application, such as spatial and temporal properties of sensors and actuators, noise-polluted sensor signal and arrangement issues. The chapter closes with a comprehensive description of the wall-sensing enabled control schemes and their influence on the control performance.

Local drag reducing control of spatially developing TBL is considered to be more important for a realistic engineering application than TCF. Chapter 4 covers the topic of locally applied reactive control in a TBL configuration. The chapter opens with a description of the numerical implementation and simulation configuration followed by a comparison of control performance in TCF and TBL. Further, differences between the control applied in a channel flow and spatially developing TBL are analysed in order to clarify the drag reducing mechanism in these flow configurations. Finally, an investigation of the downstream behaviour of the locally controlled flow is presented.

Chapter 5 summarises the novel insights drawn from the present work and provides an outlook for further possible investigations aimed at advancing numerical turbulent control schemes towards more practical application.

2 Control of Turbulent Flows – Fundamentals

2.1 Governing Equations of Fluid Dynamics

Assuming a continuum hypothesis, the fluid motion is governed by the Navier–Stokes equations [3, 104, 108]. Throughout the present work flows with a constant density ρ are considered. For an incompressible flow the continuity equation is defined as

$$\frac{\partial u_i}{\partial x_i} = 0, \tag{2.1}$$

with the velocity field denoted by $u_i(x_j, t)$, where t is time and x_j is the spatial coordinate. The coordinates $x = x_1$, $y = x_2$ and $z = x_3$ correspond to the streamwise, wall-normal and spanwise directions, respectively. The velocity components in the three directions are denoted by $u = u_1$, $v = u_2$ and $w = u_3$.

For a Newtonian fluid the Navier–Stokes equations are given by

$$\frac{\partial u_i}{\partial t} + u_j \frac{\partial u_i}{\partial x_j} = \frac{\mu}{\rho} \frac{\partial^2 u_i}{\partial x_j \partial x_j} - \frac{1}{\rho} \frac{\partial p}{\partial x_i} + f_i, \tag{2.2}$$

where p is the static pressure, μ is the dynamic viscosity and f_i is a body force per unit mass experienced by the fluid (e.g. gravity).

Typically, this set of equations is used in a non-dimensional form. The non-dimensionalisation is performed through normalisation utilising flow dependent quantities, such as characteristic velocity, u^*, and characteristic lengthscale, x^*:

$$u_i^\nabla = \frac{u_i}{u^*}, \qquad x_i^\nabla = \frac{x_i}{x^*}. \tag{2.3}$$

Based on these definitions all the remaining variables of the Navies–Stokes equations can be non-dimensionalised with

$$t^{\triangledown} = \frac{t}{x^*/u^*}, \qquad p^{\triangledown} = \frac{p}{\rho u^{*2}}, \qquad f_i^{\triangledown} = \frac{f_i}{u^{*2}/x^*}. \tag{2.4}$$

Substitution of the variables in equations 2.1 and 2.2 delivers the dimensionless continuity equation

$$\frac{\partial u_i^{\triangledown}}{\partial x_i^{\triangledown}} = 0, \tag{2.5}$$

and the dimensionless Navier–Stokes equations

$$\frac{\partial u_i^{\triangledown}}{\partial t^{\triangledown}} + u_j^{\triangledown} \frac{\partial u_i^{\triangledown}}{\partial x_j^{\triangledown}} = \frac{\nu}{u^* x^*} \frac{\partial^2 u_i^{\triangledown}}{\partial x_j^{\triangledown} \partial x_j^{\triangledown}} - \frac{\partial p^{\triangledown}}{\partial x_i^{\triangledown}} + f_i^{\triangledown}, \tag{2.6}$$

where ν is the kinematic viscosity. The reciprocal factor in the first term of the equation on the right-hand side is known as the Reynolds number:

$$\mathrm{Re} = \frac{u^* x^*}{\nu}. \tag{2.7}$$

This non-dimensional number defines the ratio between inertial and viscous forces and is used as a scaling factor for fluid dynamics problems. The Reynolds number definition is chosen based on the flow configuration. Considering fundamental flow configurations such as TCF, the Reynolds number is commonly defined using geometric dimensions of the channel and the bulk mean velocity of the flow (see Section 2.3.1). A widespread definition of the Reynolds number in TBL utilises the boundary layer thickness and the free-stream velocity of the flow (see Section 2.3.2).

2.2 Turbulent Flows and Their Statistical Description

Turbulent flows still remain some of the most difficult issues of modern physics due to the complexity of their behaviour. This complexity is linked to the multi-scale nature of turbulence and can be generally summarised through the idea of the energy cascade [116] and the Kolmogorov hypothesis [78]. The energy cascade implies a concept where the kinetic energy enters the turbulent process at the largest scale, is then transferred to the smaller scale motions and is finally dissipated at the smallest turbulent scales through viscous activity. Kolmogorov completed the theory with the quantitative definition of the smallest scales of turbulence, assuming statistical similarity and isotropy of the turbulent flows on the small scales for high Reynolds numbers. Based on the dissipation, ε, and kinematic viscosity, ν, the Kolmogorov scales are given by

$$\eta = \left(\frac{\nu^3}{\varepsilon}\right)^{\frac{1}{4}}, \qquad u_\eta = (\varepsilon\nu)^{\frac{1}{4}}, \qquad t_\eta = \left(\frac{\nu}{\varepsilon}\right)^{\frac{1}{2}}, \tag{2.8}$$

where η, u_η and t_η represent length, velocity and time scales. Using equation (2.7) and the scaling of dissipation $\varepsilon \sim {u^*}^3/x^*$ the following relationships are derived:

$$\frac{\eta}{x^*} \sim \mathrm{Re}^{-\frac{3}{4}}, \qquad \frac{u_\eta}{u^*} \sim \mathrm{Re}^{-\frac{1}{4}}, \qquad \frac{t_\eta}{t^*} \sim \mathrm{Re}^{-\frac{1}{2}}. \tag{2.9}$$

It is evident that the separation between the largest scales and the smallest scales in a turbulent flow increases with increasing Reynolds number.

The chaotic and irregular nature of turbulent motions makes it very difficult to utilise a deterministic approach for turbulence research. Thus, a statistical approach is commonly chosen and is also applied throughout the present thesis. This approach considers turbulent the velocity field, $u_i\left(x_j, t\right)$, a random variable and uses statistical tools for analysis of turbulent processes.

According to Reynolds decomposition [115], any random field $\phi(x_i, t)$ can be split into its mean and fluctuation part:

$$\phi(x_i, t) = \bar{\phi}(x_i) + \phi'(x_i, t), \tag{2.10}$$

where the mean part is defined as

$$\bar{\phi}(x_i) = \lim_{T \to \infty} \frac{1}{T} \int_0^T \phi(x_i, t)\, \mathrm{d}t. \tag{2.11}$$

Application of the decomposition to u_i, p and f_i with subsequent substitution into equation (2.6) and temporal averaging provides the Reynolds averaged Navier–Stokes equations:

$$\frac{\partial \bar{u}_i}{\partial t} + \bar{u}_j \frac{\partial \bar{u}_i}{\partial x_j} + \frac{\partial \overline{u_i' u_j'}}{\partial x_j} = \frac{1}{\mathrm{Re}} \frac{\partial^2 \bar{u}_i}{\partial x_j{}^2} - \frac{\partial \bar{p}}{\partial x_i} + \bar{f}_i, \tag{2.12}$$

while the continuity equation reads as follows:

$$\frac{\partial \bar{u}_i}{\partial x_i} = 0. \tag{2.13}$$

The newly appeared term on the left-hand side of equation (2.12) contains the second rank tensor, $\overline{u_i' u_j'}$, which is known as Reynolds stress, $\rho \overline{u_i' u_j'}$, when multiplied with ρ. The quantity represents the turbulent exchange of momentum due to the presence of velocity fluctuations.

For the characterisation of statistically stationary random processes the following definitions are used. The one-time, one-point covariance between two random variables $\phi(x_i, t)$ and $\gamma(x_i, t)$ is defined as

$$\mathrm{cov}(\phi, \gamma) = \overline{\phi'(x_i, t)\, \gamma'(x_i, t)}. \tag{2.14}$$

Hence, from the statistical point of view, the Reynolds stress appearing in equation (2.12) is the one-point, one-time covariance of the velocity. Considering identical covariance variables, the variance is defined as

$$\mathrm{var}(\phi) = \mathrm{cov}(\phi, \phi) = \overline{\phi'(x_i, t)^2}. \tag{2.15}$$

The variance is commonly used to describe the deviation of the quantity from its mean value (also known as root mean square or RMS value):

$$\phi_{\mathrm{rms}}(x_i) = \sqrt{\mathrm{var}(\phi)}. \tag{2.16}$$

The normalised autocovariance, or in other words the temporal one-point correlation coefficient between the process at time t and $t + \Delta t$, is given by

$$C_t(x_i, \Delta t) = \frac{\overline{\phi'(x_i, t)\,\phi'(x_i, t + \Delta t)}}{\overline{\phi'(x_i, t)^2}}. \tag{2.17}$$

Similarly, the spatial one-time correlation coefficient, also called the two-point correlation is

$$C_s(x_i, \Delta x_i) = \frac{\overline{\phi'(x_i, t)\,\phi'(x_i + \Delta x_i, t)}}{\sqrt{\left(\overline{\phi'(x_i, t)^2} \cdot \overline{\phi'(x_i + \Delta x_i, t)^2}\right)}}, \tag{2.18}$$

and the spatio-temporal correlation is expressed as follows:

$$C_{st}(x_i, \Delta x_i, \Delta t) = \frac{\overline{\phi'(x_i, t)\,\phi'(x_i + \Delta x_i, t + \Delta t)}}{\sqrt{\left(\overline{\phi'(x_i, t)^2} \cdot \overline{\phi'(x_i + \Delta x_i, t + \Delta t)^2}\right)}}. \tag{2.19}$$

A correlation coefficient of 1 or -1 represents the existence of a total or direct dependence between quantities considered. A correlation coefficient of 0 implies the quantities are independent of each other.

Based on definitions (2.17) and (2.18) the integral timescale

$$L_t(x_i) = \frac{1}{C_t(x_i, 0)} \int_0^{\infty} C_t(x_i, \Delta t)\, \mathrm{d}\Delta t, \tag{2.20}$$

and integral lengthscale

$$L_s(x_i) = \frac{1}{C_s(x_i, 0)} \int_0^{\infty} C_s(x_i, \Delta x_i)\, \mathrm{d}\Delta x_i, \tag{2.21}$$

are defined.

2.3 Wall-Bounded Flows

2.3.1 Fully Developed Turbulent Channel Flow

Channel flow is one of the most common fundamental flow configurations to have been reported in the literature. In theory, this configuration considers internal flow between two parallel infinitely long walls. In terms of realistic engineering applications, TCF is similar to pipe flow or flow through, for example, a heat exchanger. Assuming a statistically fully developed flow state, statistical independence of the flow in the x- and z-directions is introduced. Therefore, this set-up can be investigated numerically with a finite domain extent and application of periodical boundary conditions in both x- and z-directions. A schematic of the numerical domain configuration with the coordinate system is depicted in Figure 2.1. The extents of the domain in streamwise and spanwise directions are L_x and L_z, respectively, while the distance between the walls is $L_y = 2\delta$ with the lower wall at $y = 0$.

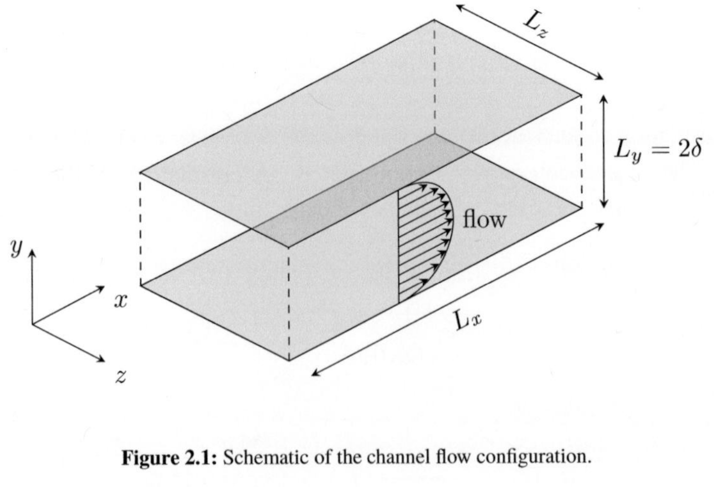

Figure 2.1: Schematic of the channel flow configuration.

The bulk Reynolds number is commonly used for the characterisation of the channel flow configuration:

$$\mathrm{Re}_b = \frac{2U_b\delta}{\nu} \quad \text{with} \quad U_b = \frac{1}{\delta}\int_0^\delta \bar{u}\mathrm{d}y. \tag{2.22}$$

Flows with $\mathrm{Re}_b > 3000$ are typically considered fully turbulent.

Considering a fully developed ($\partial\bar{u}_j/\partial x_1 = 0$) and steady ($\partial/\partial t\overline{(...)} = 0$) state along with spanwise homogeneity ($\partial/\partial x_3\overline{(...)} = 0$) and no-slip boundary conditions at the channel walls for the mean velocity and fluctuations, the Reynolds averaged mean continuity equation (2.13) is reduced to

$$\frac{\mathrm{d}\bar{v}}{\mathrm{d}y} = 0, \tag{2.23}$$

while the mean momentum equation (2.12) is transformed into

$$0 = \nu\frac{\mathrm{d}^2\bar{u}}{\mathrm{d}y^2} - \frac{\mathrm{d}\overline{u'v'}}{\mathrm{d}y} - \frac{1}{\rho}\frac{\partial\bar{p}}{\partial x} \tag{2.24}$$

for the streamwise direction and

$$0 = -\frac{\mathrm{d}\overline{v'v'}}{\mathrm{d}y} - \frac{1}{\rho}\frac{\partial\bar{p}}{\partial y} + \rho(-g), \tag{2.25}$$

for the wall-normal direction. The only conservative volume force present is gravity (acting in the negative y-direction), hence the total volume force term is assumed to be negligible.

Integration of equation (2.25) with no-slip boundary condition for $\overline{v'v'}$ at the wall gives

$$\overline{v'v'} + \frac{\bar{p}}{\rho} = \frac{p_w(x)}{\rho}, \tag{2.26}$$

where p_w is the mean pressure at the wall. The equation reveals the uniformity of the streamwise pressure gradient in TCF:

$$\frac{\partial\bar{p}}{\partial x} = \frac{\mathrm{d}p_w}{\mathrm{d}x}, \tag{2.27}$$

so that the equation (2.24) can be rewritten as

$$\frac{\mathrm{d}\tau}{\mathrm{d}y} = \frac{\mathrm{d}p_w}{\mathrm{d}x}, \tag{2.28}$$

with

$$\tau = \rho\nu\frac{\mathrm{d}\bar{u}}{\mathrm{d}y} - \rho\overline{u'v'}, \tag{2.29}$$

where τ is the total mean shear stress. Hence, the mean wall shear stress τ_w defines the total stress profile:

$$\tau(y) = \tau_w\left(1 - \frac{y}{\delta}\right). \tag{2.30}$$

Since the Reynolds stress is zero at the wall, the wall shear stress is defined as

$$\tau_w = \rho\nu\left.\frac{\mathrm{d}\bar{u}}{\mathrm{d}y}\right|_{y=0}. \tag{2.31}$$

The normalised wall shear stress is known as the skin friction coefficient

$$c_f = \frac{\tau_w}{0.5\rho U_b^2}. \tag{2.32}$$

Based on triple integration of the Navier–Stokes equations in TCF (2.24) Fukagata *et al.* [37] proposed the decomposition of the skin friction coefficient into its contributing parts, also known as the FIK-identity:

$$c_f = \underbrace{\frac{12}{\mathrm{Re}_b}}_{\substack{c_f^L \\ \text{laminar} \\ \text{contribution}}} + \underbrace{12\int_0^1 (1-y)\left(-\overline{u'v'}\right)\mathrm{d}y}_{\substack{c_f^T \\ \text{Reynolds shear stress} \\ \text{contribution}}}, \tag{2.33}$$

where y is normalised with the channel half-height δ. This division shows that c_f in the TCF consists of the laminar (c_f^L) and turbulent (c_f^T) contributions.

Considering the importance of the near-wall region, non-dimensionalisation based on the viscous units of the flow is commonly used for the channel flow configuration. The friction velocity, u_τ, and viscous lengthscale, δ_ν, are used as characteristic quantities:

$$u_\tau = \sqrt{\frac{\tau_w}{\rho}} \quad \text{and} \quad \delta_\nu = \frac{\nu}{u_\tau}. \tag{2.34}$$

The friction Reynolds number is given by

$$\mathrm{Re}_\tau = \frac{u_\tau \delta}{\nu}. \tag{2.35}$$

Using these definitions a set of equations in viscous lengths or wall units based on (2.5) and (2.6) can be derived. Quantities normalised with viscous units are denoted with the superscripted plus sign ($^+$) throughout the thesis:

$$u_i^+ = \frac{u_i}{u_\tau}, \qquad x_i^+ = \frac{x_i}{\delta_\nu} = x_i \cdot \frac{u_\tau}{\nu}, \tag{2.36}$$
$$t^+ = t \cdot \frac{u_\tau^2}{\nu}, \qquad p^+ = \frac{p}{\rho {u_\tau}^2}, \qquad f_i^+ = f_i \cdot \frac{\nu}{{u_\tau}^3}.$$

Since mean profiles in the TCF are fully specified by u_τ, δ, ν and ρ, the friction Reynolds number is widely used as the main characteristic flow parameter. For additional information on the statistical properties of TCF the reader is referred to the literature [75, 98].

2.3.2 Spatially Developing Turbulent Boundary Layer

A spatially developing turbulent boundary layer flow is a flow over a smooth flat plate with streamwise development of the boundary layer thickness $\delta_{99}(x)$. Figure 2.2 depicts the TBL flow with the coordinate system and domain dimensions, $L_x \times L_y \times L_z$. This configuration represents an external flow that is similar to the flow evolving in the vicinity of solid bounding surfaces. While TCF represents internal flow, such as in pipes and ducts, TBL is usually associated with external flow moving along bodies and surfaces or objects moving through a fluid. The main differences of the TBL in comparison with the TCF are primarily linked to the flow evolution in the streamwise direction.

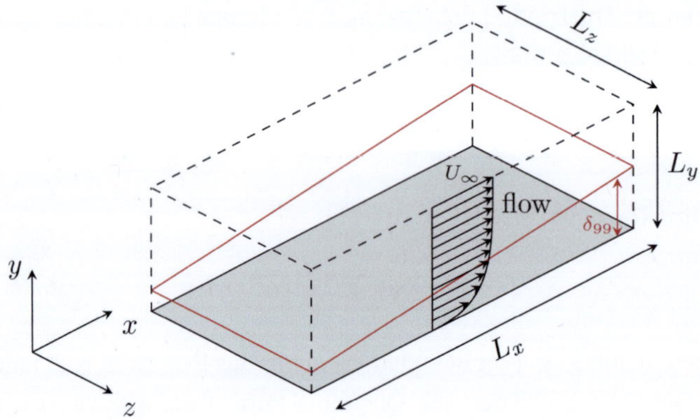

Figure 2.2: Schematic of the developing boundary layer flow.

The x-coordinate can no longer be considered statistically homogeneous, so only one periodical boundary condition in the z-direction can be applied in a numerical simulation.

The mean free stream velocity and pressure outside of the TBL are denoted by $U_\infty(x)$ and $p_\infty(x)$, respectively. Due to the absence of large velocity gradients and the insignificance of the viscosity, the outer flow is considered to be frictionless and potential. Correspondingly, based on Bernoulli's equation [108]

$$p_\infty(x) + \frac{1}{2}\rho U_\infty(x)^2 = \text{constant}, \tag{2.37}$$

the relationship between the outer pressure gradient and the gradient of the free-stream velocity is derived:

$$-\frac{\mathrm{d}p_\infty}{\mathrm{d}x} = \rho U_\infty \frac{\mathrm{d}U_\infty}{\mathrm{d}x}. \tag{2.38}$$

According to this definition a typical differentiation between accelerating flow ($\mathrm{d}U_\infty/\mathrm{d}x > 0$) with a favourable pressure gradient, decelerating flow ($\mathrm{d}U_\infty/\mathrm{d}x < 0$) with adverse pressure gradient and zero-pressure-gradient flow, where U_∞ is constant, has been established. Throughout the thesis a zero-pressure-gradient flow with flat plate at zero incidence is utilised.

The following quantities are used for the characterisation of the TBL. The boundary layer thickness, δ_{99}, is arbitrarily defined as the value of y, where $u = 0.99U_\infty$. However, due to the absence of an explicit mathematical definition, the following additional boundary layer thicknesses are proposed [126]. The displacement thickness represents the reduction of the mean volume flux:

$$\delta_d(x) = \int_0^\infty \left(1 - \frac{\bar{u}(x,y)}{U_\infty(x)}\right) \mathrm{d}y, \tag{2.39}$$

while the momentum thickness shows the reduction of the flowing momentum relative to the outer flow and is defined as

$$\theta(x) = \int_0^\infty \frac{\bar{u}(x,y)}{U_\infty(x)} \left(1 - \frac{\bar{u}(x,y)}{U_\infty(x)}\right) \mathrm{d}y. \tag{2.40}$$

Correspondingly, a variety of Reynolds numbers based on these thicknesses exist:

$$\mathrm{Re}_{\delta_{99}} = \frac{U_\infty \delta_{99}}{\nu}, \qquad \mathrm{Re}_{\delta_d} = \frac{U_\infty \delta_d}{\nu}, \qquad \mathrm{Re}_\theta = \frac{U_\infty \theta}{\nu}. \tag{2.41}$$

Furthermore, the Reynolds number defined on the basis of the streamwise coordinate is also used:

$$\mathrm{Re}_x = \frac{U_\infty x}{\nu}. \tag{2.42}$$

$\mathrm{Re}_x \approx 10^6$ is known to be the critical location, after which the TBL is considered to be fully turbulent [108]. Normalisation using viscous units is also possible using definitions (2.34) and (2.35) with boundary layer thickness, δ_{99}, instead of the channel half-height, δ, [132] so the friction Reynolds number is given by

$$\mathrm{Re}_\tau = \frac{u_\tau \delta_{99}}{\nu}. \tag{2.43}$$

According to Prandtl's approach [126], the streamwise derivatives in (2.12) are assumed to be much smaller than the wall-normal ones, and therefore can be neglected. Consequently, the well known Prandtl boundary layer equation for two-dimensional steady-flow reads as

$$\begin{aligned} \bar{u}\frac{\partial \bar{u}}{\partial x} + \bar{v}\frac{\partial \bar{u}}{\partial y} &= \nu\frac{\partial^2 \bar{u}}{\partial y^2} - \frac{\partial \overline{u'v'}}{\partial y} - \frac{1}{\rho}\frac{\partial p_\infty}{\partial x} \\ &= \frac{1}{\rho}\frac{\partial \tau}{\partial y} + U_\infty\frac{\mathrm{d}U_\infty}{\mathrm{d}x}, \end{aligned} \tag{2.44}$$

for the streamwise and

$$\frac{\partial \bar{p}}{\partial y} = 0, \tag{2.45}$$

for the wall-normal component with the continuity equation

$$\frac{\partial \bar{u}}{\partial x} + \frac{\partial \bar{v}}{\partial y} = 0. \tag{2.46}$$

The local skin friction coefficient in TBL is defined similarly to the one in TCF as the local τ_w normalised by U_∞:

$$c_f(x) = \frac{\tau_w(x)}{0.5\rho U_\infty^2}. \tag{2.47}$$

In order to describe the relationship between the streamwise evolution of TBL and its skin friction profile, the integral momentum equation based on equation (2.44) proposed by von Kármán [144] is given by

$$c_f(x) = 2\frac{\mathrm{d}\theta}{\mathrm{d}x} + \frac{4\theta + 2\delta_d}{U_\infty}\frac{\mathrm{d}U_\infty}{\mathrm{d}x}. \tag{2.48}$$

However, it has to be emphasised that this relationship is only valid for the uncontrolled laminar or turbulent flow. The simplifications introduced by Prandtl in the derivation of equation (2.44) result in an inadequate representation of the momentum balance for the TBL under adverse or favourable pressure gradients [14]. In order to achieve a valid relationship for these instances the turbulent contributions and streamwise derivatives have to be taken into account. For more details on the topic of the momentum integral equation in TBL the reader is referred here to the literature [9, 27, 120, 139, 142, 143]. A complete derivation for the two-dimensional steady-state Reynolds averaged

Navier–Stokes equations with three-dimensional turbulence can be found in the publication by Goldschmied [42] (see Appendix A.1). After minor simplifications, a more universal relationship based on the derivation is given by

$$c_f = 2\frac{\mathrm{d}\theta}{\mathrm{d}x} - \frac{2V_w}{U_\infty} - \frac{2}{\rho U_\infty^2}\int_0^\infty \frac{\partial p}{\partial x}\mathrm{d}y - \frac{2}{U_\infty^2}\int_0^\infty \frac{\partial \overline{u'u'}}{\partial x}\mathrm{d}y, \tag{2.49}$$

where V_w represents the wall-normal velocity profile at the wall. This formula is shown to be adequate for controlled and uncontrolled flows and is therefore used throughout the thesis.

A similar integral approach is used by Fukagata *et al.* [37, 66]. In contrast to TCF (equation (2.33)), the FIK-identity for TBL is given by

$$\begin{aligned} c_f(x) = \; & \underbrace{\frac{4(1-\delta_d)}{\mathrm{Re}_{\delta_{99}}}}_{\substack{c_f^\delta \\ \text{boundary layer} \\ \text{contribution}}} \underbrace{+4\int_0^1 (1-y)\left(-\overline{u'v'}\right)\mathrm{d}y}_{\substack{c_f^T \\ \text{Reynolds shear stress} \\ \text{contribution}}} \\ & \underbrace{+4\int_0^1 (1-y)(-\bar{u}\bar{v})\,\mathrm{d}y}_{\substack{c_f^C \\ \text{mean convection} \\ \text{contribution}}} \\ & \underbrace{-2\int_0^1 (1-y)^2\left(\frac{\partial \bar{u}\bar{u}}{\partial x} + \frac{\partial \overline{u'u'}}{\partial x} - \frac{1}{\mathrm{Re}_{\delta_{99}}}\frac{\partial^2 \bar{u}}{\partial x^2} + \frac{\partial \bar{p}}{\partial x}\right)\mathrm{d}y}_{\substack{c_f^D \\ \text{spatial development} \\ \text{contribution}}}, \end{aligned} \tag{2.50}$$

where δ_d represents the displacement thickness. In this equation all variables are non-dimensionalised by U_∞ and δ_{99}. The turbulent contribution, c_f^T, is obviously present for the TCF and TBL cases, while the boundary layer contribution, c_f^δ, from TBL can be compared with the laminar contribution, c_f^L, in TCF. For TBL two additional terms, namely c_f^C and c_f^D, are present. The terms are related to the spatial development of the TBL and to the mean convection in the wall-normal direction, which is not present for TCF. Even though the FIK-identity enables a detailed mathematical analysis of c_f, it does not provide a complete physical explanation for all of the contributing parts. In particular, the newly appeared terms, c_f^C and c_f^D, can present difficulties in establishing a connection between the mathematical term and its phenomenological meaning. Hence, the physics behind particular terms is a matter of interpretation.

The computation of the streamwise derivatives in equation (2.50) is linked to certain issues when applied to a set of noisy data, especially for the data acquired in experiments. In addition, for the DNS data the estimation is difficult if the statistics are not entirely converged. A small modification of the FIK-identity (2.50) proposed by Mehdi *et al.* [90, 91] can be used in order to avoid the computation of streamwise gradients:

$$c_f(x) = \underbrace{\frac{4(1-\delta_d)}{\mathrm{Re}_{\delta_{99}}}}_{\substack{c_f^{\delta} \\ \text{boundary layer} \\ \text{contribution}}} + \underbrace{4\int_0^1 (1-y)\left(-\overline{u'v'}\right)\mathrm{d}y}_{\substack{c_f^{T} \\ \text{Reynolds shear stress} \\ \text{contribution}}} \tag{2.51}$$

$$+ \underbrace{2\int_0^1 (1-y)^2\left(-\frac{\partial\tau}{\partial y}\right)\mathrm{d}y}_{\substack{c_f^{C} + c_f^{D} \\ \text{mean convection and spatial} \\ \text{development contribution}}},$$

with the total shear stress, τ, as defined in (2.29). According to this expression, the last term represents additional contributions that appear only in TBL.

2.4 Flow Control

> Boundary layer control includes any mechanism or process through which the boundary layer of a fluid flow is caused to behave differently than it normally would were the flow developing naturally along a smooth straight surface. J. Flatt [30]

The history of flow control began with Prandtl in 1904 with the application of suction on a cylinder surface producing a delay in the boundary layer transition [109]. Nowadays, flow control techniques are moving towards microstructures and microelectromechanical systems (MEMS) aimed at the modification of the turbulent properties of the flow in order to achieve better technological performance with less environmental impacts. The present thesis is focused on the control techniques addressing the reduction of the skin friction drag in turbulent flows.

According to Gad-el-Hak [40], flow control strategies can be divided into passive control, which does not require any input power to apply the control, and active control, which corresponds to a control type where a device requiring energy expenditure is utilised. Flow additives [12, 127, 140], modification of the surface geometries using riblets [4, 5, 20, 146] and superhydrophobic surfaces [24, 38, 106, 121] account for the most prominent examples of the passive flow control technique. Active flow control implies a control loop, which can be further divided into predetermined control and reactive control. Predetermined control applies energy without receiving any information about the particular flow state and can, therefore, also be characterised as an open-loop control. In contrast, the closed-loop or reactive control is a type of control where the control input is continuously adjusted by using sensor information. Hence, the control type provides an energy input that is appropriate for the current flow state. Finally, the reactive flow control can be classified, according to the measured quantities, into feed-forward and feed-back control. The measured and controlled variables differ in the feed-forward concept, while the feed-back concept requires a measurement of the controlled variable only.

Considering a quasi-deterministic approach instead of a stochastic approach, the reduction of the turbulent skin friction drag is commonly related to the modification of the near-wall structures of the turbulent flow [16]. The existence of coherent turbulent motions was identified in the late 1970s [11] and has subsequently been investigated over several decades. It is known that the enhancement of the turbulent skin friction drag in comparison to the laminar one is linked to the presence of quasi-streamwise vortices (QSV). QSV are coherent structures with a diameter of approximately 20 – 50 viscous units appearing in the near-wall region of the turbulent flow [71]. The streamwise elongation of the vortical structures is about 400 viscous units [58]. The vortices cause upward flow motion raising the low momentum fluid from the near-wall region into the outer flow, known as ejections, as well as downward flow motion pushing high-momentum fluid towards the wall, known as sweeps. The latter events significantly contribute to the increase in the skin friction drag due to the induced steeper gradient of the streamwise velocity at the wall. A summary of the nature of coherent turbulent structures can be found in [52, 117]. Exploiting the knowledge of coherent structures and their importance, a broad variety of active control methods aimed at the reduction of skin friction drag in turbulent wall-bounded flows has been introduced in the past. Classical active control schemes, such as opposition control [18], suboptimal and optimal control techniques [8, 17, 83], wall oscillation or deformation methods [26, 64] and direct damping of near-wall fluctuations [33, 54], have been thoroughly investigated over the last 20 years. A comprehensive summary of the topic of practical applications of turbulence control aimed at the skin friction drag reduction can be found in [40, 69, 133].

Since the modification of the near-wall region of the flow field is essential for the control application, control activation is typically performed for $y \leq y_c$. Hence, as depicted in Figure 2.3, the control distribution function is defined as

$$d(x, y) = \begin{cases} 1, & \text{for} \quad x_0 \leq x \leq x_0 + \Delta x_c \\ & \text{and} \quad 0 \leq y \leq \Delta y_c \\ 0, & \text{otherwise,} \end{cases} \tag{2.52}$$

where x_0 represents the streamwise placement of the control volume with extents $\Delta x_c \times \Delta y_c$ in the x- and y-directions. A control with non-uniform distribution in the x-direction (i.e., one that does not cover the entire simulation domain) is referred to as partial control throughout the thesis. In the z-direction the control volume entirely covers the domain, so the distribution function, d, does not depend on the z-coordinate. For the control methods where only wall actuation is utilised, the distribution function is only given for $y = 0$. In general, the majority of control techniques are first tested in a TCF with the control application to the entire wall area.

The present investigation considers various active control schemes, whereby the focus is mainly on the reactive control application. The following drag reducing control schemes are examined in this thesis: local blowing, opposition control, suboptimal control and body force damping. The control performance indices will be introduced in the following section, after which an overview of the considered control techniques will be presented in Sections 2.4.3–2.4.2.

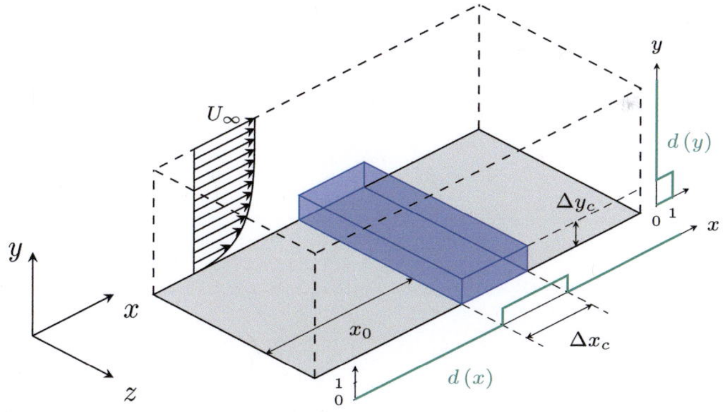

Figure 2.3: Definition of the control placement in the flow field.

2.4.1 Control Performance Indicies

A modification of the flow field in terms of skin friction drag due to the application of flow control has to be quantified through the definition of several control performance indices. When considering TCF, the following definitions are applied [69]. With respect to the uncontrolled case, the reduction rate of skin friction drag is given by

$$R = 1 - \frac{c_f}{c_{f,0}}, \tag{2.53}$$

where the subscript "0" denotes the uncontrolled value. If the flow rate in a channel flow is kept constant (CFR), the modification of the skin friction coefficient is reflected in a change in τ_w or u_τ:

$$R = 1 - \frac{\tau_w}{\tau_{w,0}} = 1 - \left(\frac{u_\tau}{u_{\tau,0}}\right)^2, \tag{2.54}$$

or utilising the definition of pumping power in TCF

$$P = 2U_b\tau_w, \tag{2.55}$$

the drag reduction rate is alternatively given by

$$R = 1 - \frac{P}{P_0}. \tag{2.56}$$

For a constant pressure gradient condition (CPG), where the wall shear stress remains constant, control activation leads to an increase in U_b, so the control effect is measured by

$$R = 1 - \frac{c_f}{c_{f,0}} = 1 - \frac{U_b^2}{U_{b,0}^2}. \tag{2.57}$$

These definitions can be applied to passive or active flow control techniques. In active control, additional energy expenditure or control input power, P_{in}, is required. Hence, the energy saving rate, S, replaces R by taking into account P_{in}, as in

$$S = 1 - \frac{(P + P_{in})}{P_0}. \tag{2.58}$$

Positive S represents an energetically beneficial control application. The energy gain is defined as the ratio of the reduced pumping power to the applied control power input:

$$G = \frac{(P_0 - P)}{P_{in}}. \tag{2.59}$$

Energy gain is another control performance index that shows whether the control application is advantageous in terms of energy balance. $G < 1$ corresponds to the case when energy expenditure caused by control activation is higher than the energy saved due to the control, regardless of the drag reduction rate. Hence, $G > 1$ is needed for an energy efficient control application.

Throughout the thesis we assume 100% actuator efficiency, so only the ideal fluid mechanical power input is considered. In a real application scenario an actuator efficiency of < 100% has to be taken into account, which definitely leads to lower achievable values for S and G.

The control power input strongly depends on the actuation type. In the case of the wall-normal velocity applied at the wall surface, the ideal amount of energy required for the control activation can be defined as the energy flux through the walls into the flow system caused by the imposed velocity v:

$$P_{in,v} = \left(|\overline{pv}| + \frac{1}{2}|\overline{\rho v^3}| \right)_{y=0}. \tag{2.60}$$

Accordingly, the energy consumption required for the velocity imposed at the wall surface in the spanwise direction can be estimated as work of the spanwise shear force by

$$P_{in,w} = \mu \left(\left| \overline{w \frac{dw}{dy}} \right| \right)_{y=0}, \tag{2.61}$$

where μ^* is the fluid viscosity. For the control schemes, where flow control is realised through application of a body force, the control power is estimated as

$$P_{in,f} = \left| \overline{f_i u_i} \right|, \tag{2.62}$$

where f_i is the imposed body force.

The indices introduced here consider local values per unit area, so in TCF, integration in the statistically homogeneous directions x and z at the wall surface has to be performed. A triple integration has to be carried out within the volume where the body force is applied for the calculation of power input in equation (2.62). It should also be noted that the exact mathematical form of the power input from actuators to the fluid system is given by equations (2.60) – (2.62) without taking the absolute value of each term. Since the control input power is strictly dependent on the actuator type and implementation details, it is difficult to propose a universal definition for it. We consider the most conservative method of control input power characterisation using the absolute values for the terms in equations (2.60) – (2.62) assuming that introduction of any control input is always linked to some energy expenditure, independent of the direction of the imposed velocity or body force. This scenario represents the "worst case" energy consumption from the fluid dynamical point of view.

Similarly, the control performance indices are introduced in TBL using U_∞ instead of U_b, so the local driving power is given as

$$P(x) = U_\infty(x)\,\tau_w(x)\,. \tag{2.63}$$

Hence, the identical definitions (2.54) – (2.59) for the performance description as well as (2.60) and (2.61) for the estimation of control input power are used in TBL. However, due to the streamwise development of the flow only the integration in the z-direction is utilised (z- and y-directions for body force control input), so all the indices are defined per unit length and are dependent on the x-coordinate.

2.4.2 Reactive Control

Reactive control, which is also referred to as feedback control, operates actuators based on the instantaneous flow information obtained by sensors. Thus, it generally enables flexible control of turbulence and offers large energy gains with low consumptions of control power. The schemes utilise blowing and suction at the wall surface and are mainly designed with the aim of Reynolds shear stress suppression and hence reduction of the near-wall turbulent activity.

Opposition Control

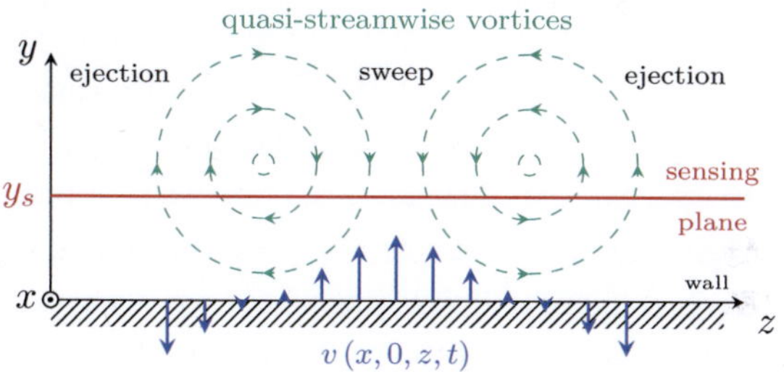

Figure 2.4: Schematic of the wall-normal opposition control.

Opposition control is one of the classical reactive control schemes, introduced by Choi *et al.* in 1994 [18]. Control activation is performed by local suction and blowing in the wall-normal direction at the wall surface, so as to suppress the sweep and ejection events in the near-wall region and reduce the skin friction drag. In TCF the control is commonly applied to the entire area of the wall, imposing wall-normal or spanwise velocity opposite to the velocity captured at a prescribed sensing plane y_s, as illustrated in Figure 2.4, where a cross-section for an x-coordinate within the control region is shown. Correspondingly, the wall-normal control input at the wall is given by

$$v(x, 0, z, t) = -\alpha \cdot d(x) \cdot v(x, y_s, z, t), \tag{2.64}$$

while the spanwise control is defined as

$$w(x, 0, z, t) = -\alpha \cdot d(x) \cdot w(x, y_s, z, t), \tag{2.65}$$

where α is a positive amplification factor and the control placement is defined by $d(x)$. The original paper [18] reports $R = 25\%$ and $R = 30\%$ at $\mathrm{Re}_\tau = 180$

with entire wall actuation for wall-normal and spanwise opposition control, respectively. A relaminarisation of the flow is achieved for $\mathrm{Re}_\tau = 100$, corresponding to $R = 63\%$. This work points out the dependency of control performance on the position of the sensing plane, y_s, declaring $y_s^+ = 10$ as the optimal position. A subsequent investigation by Hammond *et al.* [47] shows that the sensing plane should be placed further from the wall at $y_s^+ = 15$ for better control performance. A Reynolds number dependency of the optimal sensing plane position is established in [15, 53], showing a decrease in the optimal position in viscous units with increasing Reynolds number. The influence of interactions between the amplification α and the sensing plane position on the control performance is investigated in [23]. According to this study, a sensing plane position above or below the optimal one always leads to a worse control performance, but the effect can be lessened by an adjustment of α. The opposition control scheme shows similar results when applied in a turbulent pipe-flow [34] or in TBL [102].

The mechanism of opposition control can clearly be seen in the comparison of statistical flow properties presented in Figure 2.5. The wall-normal coordinate is plotted in logarithmic scale in order to highlight the near-wall region when control is introduced. The controlled flow shows a strong attenuation for u^+_{rms} and w^+_{rms} throughout the entire domain height, while the fluctuation activity in the wall-normal direction, v^+_{rms}, and Reynolds shear stress are significantly reduced for $y^+ > 5$. The imposed velocity is noticeable in the near-wall region with $v^+_{rms} = 0.19$ at the wall. One can also recognise the so-called virtual wall in the profile of v^+_{rms} at $y^+ = 8$, marking the position where v-fluctuations almost vanish. The behaviour of the flow at this position is similar to the one at the wall of an uncontrolled TCF. The actuation also influences the near-wall Reynolds shear stress for $y^+ < 5$, where non-zero values are observed. However, the total shear stress profile is significantly reduced, so the wall shear stress $\bar{\tau}_w = 0.763$ corresponding to $R = 23.7\%$ is observed. Considering componental decomposition of c_f from equation (2.33) demonstrated in Figure 2.6, one can confirm that the reduction of the skin friction drag for opposition control is entirely attributed to the reduction of the Reynolds shear stress contribution, c_f^T, while the laminar contribution, c_f^L, remains unaffected.

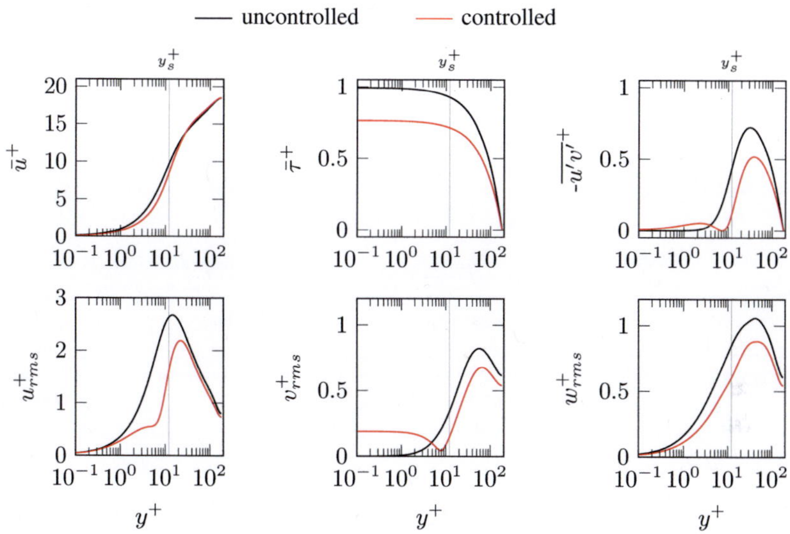

Figure 2.5: Change of the statistical flow properties due to the application of wall-normal opposition control in TCF at $\mathrm{Re}_\tau = 180$ with sensing plane position $y_s^+ = 12$ resulting in $R = 23.7\%$.

Several works on the realisability of reactive control techniques similar to opposition control can be found in the literature [48, 69, 86], which suggest MEMS devices for a practical implementation. However, some workers also point out various issues linked to the MEMS fabrication process, the problematic nature of sensing and inefficient control algorithms. To the best of the author's knowledge there is only one experimental study on the basis of opposition control, which was conducted by Rebbeck & Choi [113]. The study confirms the possibility of the near-wall flow structure detection and their attenuation by application of selective blowing.

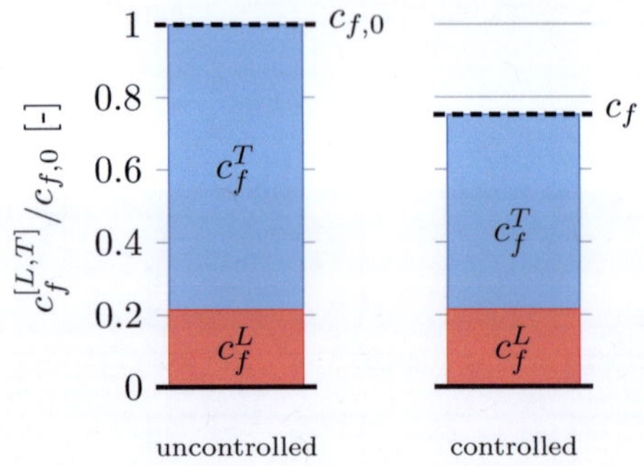

Figure 2.6: FIK-decomposition of c_f for uncontrolled and controlled case at $\mathrm{Re}_\tau = 180$. Normalisation is performed with the skin friction of the uncontrolled flow, $c_{f,0}$.

Based on opposition control, a turbulence control method using wall deformation is developed [26, 68, 94]. The researchers report a deterioration of the drag reducing effect to 10 – 17% at $\mathrm{Re}_\tau = 150$ due to the limitation introduced by wall-deformation amplitude. Pamiès *et al.* [103] investigates a more realistic control set-up with turbulent-structure-like wall-deformation actuators using DNS. The achieved drag reduction does not exceed 6%.

Suboptimal Control

In 1990 Abergel & Temam [1] proposed a mathematical description of the reactive scheme called optimal control. The scheme utilises the entire velocity and pressure information in the simulation domain over a certain period of simulation time in order to estimate the optimal control input and achieve drag reduction. The control input is derived analytically based on the Navier–Stokes equations in a framework of a TCF driven by a constant pressure gradient. Bewley *et al.* [8] showed that optimal control theory provides high efficiency resulting in $R = 50\%$ at $\mathrm{Re}_\tau = 180$ in TCF. However, the method implies high computational costs due to the long time horizon involved in the estimation of the control input.

The scheme of suboptimal control introduced by Choi *et al.* [17] can be considered as a modification of the optimal control scheme. The suboptimal control scheme reduces the time horizon for optimum seeking down to a simulation time step and therefore cannot provide a global optimum value. These researchers use $(\partial u/\partial x)^2$ for the definition of cost function and derive the control input based on the Burgers equation [13] as body force or blowing and suction at the wall surface. The reduction of the cost function is reported to be up to 87%. Bewley *et al.* [7] derived the control input in the form of blowing and suction at the wall in a channel flow at $\mathrm{Re}_\tau = 100$. The control application yields $R = 17\%$. Lee *et al.* [83] applied suboptimal control based on linearised Navier–Stokes equations using $\partial w/\partial y$ or pressure at the wall as sensing information. Skin friction drag reduction of $R = 16 - 22\%$ was achieved in a TCF at $\mathrm{Re}_\tau = 110$. When only wall-based quantities are used as input for the suboptimal control scheme, the main disadvantage of opposition control, namely sensor placement inside the flow, can be overcome.

The scheme of suboptimal control can also be adopted for the configuration of the flow around bluff bodies [19, 93]. Jeon & Choi [56] implemented a suboptimal control scheme aimed at drag reduction in a configuration of the flow over a sphere. A modification of the control scheme is performed, so that the weight function is computed numerically once at the start of the simulation in order to be able to estimate the control input. While earlier examples of suboptimal control application are restricted to a specific flow configuration or boundary condition, this method is more general and can be employed in different geometrical configurations. Using surface pressure as sensor information and the pressure deviation from the potential flow solution as a cost function, the authors reported $R \approx 30\%$ for $\mathrm{Re} = 425$ (based on sphere diameter and free-stream velocity).

The opposition control and suboptimal control schemes are closely related. It is known that the velocity distribution estimated by suboptimal control resembles the control input provided by opposition control [7]. During the derivation of the suboptimal control performed by Lee *et al.* [83] observations of opposition control results are used for the definition of the cost function, so the control is designed by mimicking opposition control. This undermines the assumption of the similar nature of the control effect for opposition control and suboptimal control, namely damping of the near-wall turbulent structures. However, the reported drag reduction is slightly lower in suboptimal control than in opposition control. This can be explained by the placement of the sensors: while in opposition control the sensors directly capture the information about the near-wall structures, suboptimal control has to estimate actual velocity distributions based on wall information. In terms of statistical properties, the flow controlled with suboptimal control is very similar to the flow with applied opposition control (see Figures 2.5 and 2.6).

Body Force Damping

The scheme of body force damping utilises volume forces for the modification of the flow. The reactive scheme is introduced by Satake & Kasagi [122] for the damping of the spanwise velocity fluctuations and further investigated by Lee & Kim [81] and Iwamoto *et al.* [54] for TCF at different Reynolds numbers. Similarly to opposition control, the control law aims at the suppression of turbulent fluctuations in the near-wall region and uses velocity as the sensor information. The control input is given in the form of a body force in all three directions for a damping layer with thickness Δy_c:

$$f_i(x, y, z, t) = -\frac{d(x, y)}{\Phi} \cdot u_i(x, y, z, t), \tag{2.66}$$

with the forcing time constant Φ and distribution $d\,(x, y)$ defined in equation (2.52). A drag reduction rate of 35% in a TCF at $\mathrm{Re}_\tau = 110$ was reported by Lee & Kim [81] when the control is applied for the damping of spanwise velocity fluctuations. As shown by Iwamoto *et al.* [54], this technique provides $R = 60 - 74\%$ at $\mathrm{Re}_\tau = 642$ depending on the damping layer thickness $(10 \leq \Delta y_c^+ \leq 60)$ if the control is applied to all three components of the velocity fluctuations. Subsequently, an investigation by Frohnapfel *et al.* [33] showed application of the scheme using only the spanwise component for sensing and actuation such that

$$f_z(x, y, z, t) = -\frac{d(x, y)}{\Phi} \cdot w(x, y, z, t). \tag{2.67}$$

Control application in TCF within a damping layer of only $\Delta y_c^+ = 5$ results in $R = 29\%$ at $\mathrm{Re}_\tau = 150$, while $\Delta y_c^+ = 30$ leads to the relaminarisation of the flow field. Workers have also proposed a control scheme based on the sensing of the streamwise wall shear stress yielding $S \approx 20\%$ for several configurations at $\mathrm{Re}_\tau = 150$.

2.4.3 Predetermined Control

The majority of predetermined drag reducing control schemes in TCF involve an introduction of prescribed velocity at the wall surface mimicking wall movement [110]. In TBL drag reduction can also be achieved by an introduction of a wall-normal mass flux through the wall or blowing, which is a simple and efficient alternative to the more complex reactive control schemes.

Uniform Blowing or Suction

The most prominent example of a flow manipulation is the local suction or blowing at the wall of a flat plate boundary layer. The control scheme does not utilise any information about the instantaneous flow field and thus can be classified as a predetermined active control technique. The control can be imagined to be implemented in reality by transpiration through a porous wall or by direct suction or blowing through a slot on the wall surface. Figure 2.7 shows a schematic of the locally applied blowing in the x-y-cross section of the simulation domain. The wall-normal velocity profile at the wall is given by

$$v\,(x, y = 0, z, t) = V_w \cdot d\,(x), \tag{2.68}$$

where V_w is the velocity amplitude and $d\,(x)$ is the function defining the local control placement (see equation (2.52)).

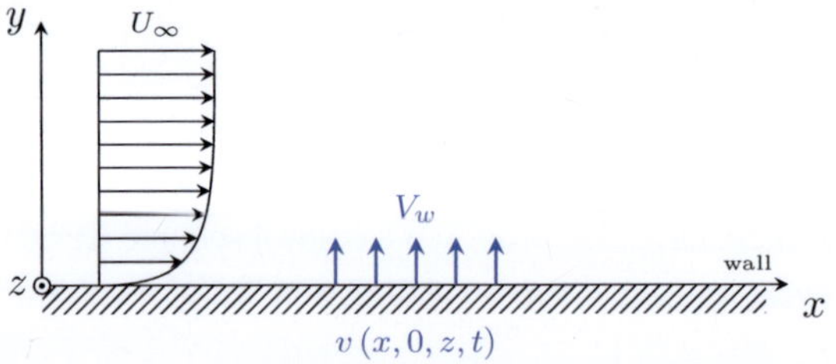

Figure 2.7: Schematic of local blowing applied to a boundary layer flow.

The topic of blowing and suction has been addressed analytically and experimentally in a series of publications beginning in the 1950s and 1960s where theoretical descriptions of the local blowing or suction on a flat plate are summarised and the first measurements were conducted [10, 92, 119]. The estimation of the control effect on skin friction coefficient based on the von Kármán momentum integral equation (2.48) demonstrates a good agreement with experimental data [77]. It was shown that local injection into the turbulent flow field causes a decrease of the skin friction drag while local suction of the fluid induces an opposite effect. The observations attest to a strongly pronounced impact on the boundary layer development even for low blowing or suction magnitude of less than 1% of the free-stream velocity. Towards the end of 1990s, the investigations in this area continued using numerical simulations. Park and Choi [105] conducted a direct numerical simulation of a TBL with local uniform blowing or suction. This work confirms previously conducted experiments showing a rapid decrease of skin friction coefficient in the blowing section with subsequent enhancement of turbulence activity downstream. A physical explanation for the effect is found when turbulent structures are observed: blowing shifts turbulent structures to the outer flow, while suction draws structures towards the wall and increases viscous diffusion. It is shown that the reduction of skin friction drag caused by blowing is followed by skin friction increase due to flow destabilisation. The paper by Kim *et al.* [76] emphasises the influence of blowing/suction area length on the downstream pressure fluctuations. Kametani and Fukagata [66] investigated various amplitudes of uniform blowing and suction and reported drag reduction of up to 80% in a TBL for blowing with 1% of U_{∞}.

An investigation of a blowing-only or a suction-only case is difficult in TCF, since the control application introduces additional mass flux and therefore a change in U_b. As a work-around blowing can be applied on one wall and compensating suction on another [21, 107, 137]. Effects similar to the TBL investigations have been reported: blowing reduces wall shear stress and increases turbulent stresses, while suction leads to the opposite effect. Although blowing applied in TBL is shown to be a simple and efficient drag reducing flow control technique, several questions such as control influence on the wall-normal momentum conservation and the flow development in the downstream of the control are still to be addressed.

3 Control of Fully Developed Turbulent Channel Flows

Fully developed turbulent channel flow is the most widely used configuration for the investigation of turbulence control in wall-bounded flows. The popularity is due to the possibility of performing a DNS with comparatively low computational costs, as a consequence of the periodicity of the flow in the streamwise and the spanwise directions [75, 95, 97]. At the same time, high spatial and temporal resolution of such TCF DNS ensures a complete representation of the turbulent processes, so the validity of the resulting controlled solution is typically not in doubt. Although a significant drag reduction rate can be obtained by the application of various active and reactive control schemes in DNS, application of such control in experiments poses new challenges of achieving similar control effects. The present chapter addresses the topic of realistic limitations of reactive control application in the framework of a TCF.

Considering a comparison of the control application in a numercal simulation with a general realistic experimental set-up, the following essential limitations arise. One of the most fundamental issues is the spatial resolution of sensors and actuators. So far, most control algorithms developed in DNS assume the unlimited spatial resolution of sensing and actuation. On the other hand, the smallest available devices still exhibit dimensions of approximately 200μm [69]. The spatial resolution of sensing and actuation is particularly crucial in high Reynolds number flows, where the physical dimension of near-wall turbulent structures becomes quite small (O(10μm)). Endo *et al.* [26] were the first to take into account the finite size of a wall deformation actuator in their DNS at a low Reynolds number. Later, Fukagata *et al.* [39] discussed the effect of the spatial resolution of actuation at a higher Reynolds number by damping either the small-scale or large-scale wall-normal velocity fluctuation via actuation. When the large-scale fluctuations are suppressed by the control, the small-scale fluctuations are found to be drastically enhanced. This implies that insufficient spatial resolution causes significant deterioration of the control performance.

Another major equipment-related restriction of an experimental set-up is the temporal resolution of the sensors and the frequency response of the actuators. Modern state-of-the-art microelectromechanical systems (MEMS) provide sensor frequency response of up to 4kHz, while the actuator frequency response is limited to 10kHz [69]. However, the effect of limited temporal resolution of sensing and actuation on the control performance has not yet been fully explored. In addition, it is highly possible that the sensor information is significantly contaminated by external noise in real systems. This is also a topic that has as yet received little attention.

A further important limitation refers to the sensing quantity employed in a control loop. Since sensing in the midst of a flow is quite difficult in real systems, several feedback control algorithms based only on wall information have been developed [79, 82, 83]. In these studies, the wall shear stress or the wall pressure is used as a sensing quantity. However, these quantities are still difficult to measure even with small-scale array sensors, and only the measurement of the streamwise wall shear stress is considered to be feasible [69]. Accordingly, control algorithms based on the streamwise wall shear stress have also been developed [33, 36].

Clearly the list of the considered limitations is not complete and could be extended significantly depending on the particular application case. For instance, the actuation type (volume force, blowing/suction) or specific actuation implementation provides additional limitations regarding the distribution and shape of the possible flow modifications, time lag in the actuation response or blockage issues. However, the chosen limitations apply to the entire range of real-world actuation and sensing conditions and therefore can be considered universal.

In Section 3.1, the numerical method, the TCF flow conditions, the control algorithms and their precise configurations are reported. Section 3.2 deals with the influence of the spatial properties of the control on the achievable control performance. The effects of the limitations linked to the temporal sensing resolution are investigated in Section 3.3. Then in Section 3.4 a random white noise with prescribed amplitude is superimposed onto the sensing quantity in order to explore the effects of noise on the control performance. Finally, Section 3.5 investigates various possibilities of reactive control application utilising wall sensing. The section describes the derivation of the reactive scheme with wall sensing based on opposition control considering the correlation approach (Section 3.5.1), the approach based of the impulse response of the TCF (Section 3.5.2) and the suboptimal control scheme with wall sensing (Section 3.5.3).

The investigations discussed Sections 3.3 and 3.4 have been previously published in [134], so the sections contain paraphrased passages and direct quotations from the publication (highlighted with quotation marks).

3.1 Numerical Procedure

The investigation of control application is performed through DNS of a TCF under CFR unless indicated otherwise. The Navier–Stokes equations are numerically integrated by a finite difference method (FDM) on a staggered grid with a fractional step method [74] for pressure decoupling. For temporal advancement, the convection and viscous terms are discretised using the 2nd order Adams–Bashforth and Crank–Nicholson methods, respectively. The flow is bounded by the upper ($y = 0$) and lower wall ($y = 2\delta = 2\mathrm{Re}_\tau$), while the spanwise and streamwise boundary conditions are periodic. At the top and bottom walls, no-slip conditions are applied except when a velocity component is imposed by control. Various simulation set-ups are investigated within a Reynolds number range of $\mathrm{Re}_\tau = 150 - 664$. A summary of utilised configurations is presented in Table 3.1. It is well known that finite difference schemes provide lower accuracy (mainly due to higher levels of differentiation error) compared to spectral methods and hence require roughly twice as high spatial resolution in all three direction in order to achieve the same results [96]. Aware of the fact that insufficient resolution can lead to a deviation in shear stress and higher order statistics, a rather coarse grid is chosen for simulations in the present investigation due to the large amount of control parameters, which can eventually cause immense overall computational costs. The appropriate simulation configuration is chosen based on the range of the parametric study and availability of computational resources. Nevertheless, all simulation configurations are validated and show a good agreement with literature data [50, 53, 75].

Due to the requirements of the applied numerical schemes the simulation time step is $\Delta t_0^+ = 0.01 - 0.03$, so the global maximal Courant–Friedrichs–Lewy condition (CFL) does not exceed 0.2. Minimal simulation time for the statistical integration is $\Delta t^+ = 4000$, where the initial temporal transient after control activation is always excluded from the statistical integration. This integration time approximately corresponds to $10 - 20$ eddy turnovers based on $\delta = \mathrm{Re}_\tau$ and u_τ. The statistics of the uncontrolled reference cases are in good agreement with literature data [50, 53, 98].

		grid nodes	domain size		resolution	
#	Re_τ	$(N_x \times N_y \times N_z)$	$(L_x^+ \times L_y^+ \times L_z^+)$	Δx^+	Δy^+	Δz^+
1	150	$64 \times 129 \times 64$	$2.5\pi\mathrm{Re}_\tau \times 2\mathrm{Re}_\tau \times \pi\mathrm{Re}_\tau$	18.7	0.1-5.7	7.4
2	150	$120 \times 129 \times 120$	$2.5\pi\mathrm{Re}_\tau \times 2\mathrm{Re}_\tau \times \pi\mathrm{Re}_\tau$	9.8	0.1-5.7	3.9
3	150	$128 \times 129 \times 128$	$2.5\pi\mathrm{Re}_\tau \times 2\mathrm{Re}_\tau \times \pi\mathrm{Re}_\tau$	9.2	0.1-5.7	3.7
4	180	$128 \times 129 \times 128$	$2.5\pi\mathrm{Re}_\tau \times 2\mathrm{Re}_\tau \times \pi\mathrm{Re}_\tau$	11.0	0.1-6.9	4.4
5	200	$128 \times 129 \times 128$	$2.5\pi\mathrm{Re}_\tau \times 2\mathrm{Re}_\tau \times \pi\mathrm{Re}_\tau$	12.3	0.1-7.7	4.9
6	300	$160 \times 257 \times 128$	$2\pi\mathrm{Re}_\tau \times 2\mathrm{Re}_\tau \times \pi\mathrm{Re}_\tau$	11.9	0.1-5.7	7.4
7	500	$256 \times 353 \times 192$	$2\pi\mathrm{Re}_\tau \times 2\mathrm{Re}_\tau \times \pi\mathrm{Re}_\tau$	12.3	0.2-6.9	8.2
8	664	$256 \times 353 \times 192$	$2\pi\mathrm{Re}_\tau \times 2\mathrm{Re}_\tau \times \pi\mathrm{Re}_\tau$	16.3	0.3-9.2	11

Table 3.1: Properties of the considered simulation domains.

The size of the numerical domain plays a major role in the maintenance of a proper fully developed turbulent state of the flow. Jiménez & Moin [58] proposed a minimal flow unit that is required for regeneration and preservation of a turbulent flow state in TCF. They suggest a minimal domain size of $(L_x^+ \times L_z^+) = (250 - 350 \times 85 - 110)$, which is directly linked to the dimensions of near-wall turbulent structures and the spacings between them. Further reduction of the domain size leads to an erroneous estimation of various flow properties in the simulation due to the fact that simulation domain cannot adequately accommodate turbulent structures and periodic boundary conditions in the x- and z-directions contribute to the development of a non-physical flow state within the domain. For a domain size smaller than the minimal flow unit $((L_x^+ \times L_z^+) < (250 \times 100))$, the self-sustaining mechanism of wall turbulence is interrupted and a relaminarisation of the flow field occurs. Further investigations on the issue of the domain size can be found in [31, 59–61, 87]. Within the framework of a controlled TCF the concern about an appropriate domain size can become even more critical, since the control application in a CFR TCF attenuates the turbulent activity and thus increases the viscous length scale. This increase effectively reduces the domain dimensions in viscous units, so a drag reducing control application in CFR TCF naturally tends to suffer from this issue.

In order to clarify the effect of the domain size on the estimated drag reduction rate, a series of TCF simulations with variations in the domain dimensions is carried out. It is decided to perform TCF simulations where flow is driven by a CPG condition in order to be able to maintain constant Re_τ and avoid variation in the domain size in viscous units for the controlled case compared with the uncontrolled one. A detailed overview of the procedure and results can be found in Appendix A.2. Based on the results it is concluded that the minimal streamwise and spanwise domain extent chosen for an investigation of opposition control in CPG TCF should not be smaller than 600 and 250 viscous units corresponding to a domain size of approximately $(L_x^+ \times L_y^+ \times L_z^+) = (3\mathrm{Re}_\tau \times 2\mathrm{Re}_\tau \times 1.25\mathrm{Re}_\tau)$. Considering controlled CFR TCF, where viscous scales change due to control application, it is advisable to maintain this minimal domain size with respect to viscous Reynolds number of the controlled flow. In this case, the appropriate minimal domain size is linked to the achievable drag reduction rate:

$$\frac{\mathrm{Re}_\tau}{\mathrm{Re}_{\tau,0}} = \sqrt{1-R}, \tag{3.1}$$

and the domain size should be consequently scaled by the factor $\frac{1}{\sqrt{1-R}}$. However, since R is not known *a priori*, an exact estimation is difficult. For application of opposition control with $R \approx 25\%$ the factor is 1.8 and the domain size can be represented as $(L_x^+ \times L_y^+ \times L_z^+) = (5.5\mathrm{Re}_{\tau,0} \times 2\mathrm{Re}_{\tau,0} \times 1.75\mathrm{Re}_{\tau,0})$ with respect to the viscous Reynolds number of the uncontrolled solution. All considered CPG and CFR TCF simulations in the present thesis fulfil the proposed domain size limitation.

3.2 Sensing and Actuation with Reduced Resolution

Figure 3.1 shows the schematic of a classical opposition control scheme, where the sensing plane is located above the wall at position y_s. Typically, control schemes proposed for DNS are applied in such a way that every grid node can act as a sensor or an actuator. For a real application this would imply unrealistically small actuators and high-resolution sensor information being available at any position in the flow domain. Thus, an investigation of the influence on the control performance linked to the actuator size and the spatial coarsening of the sensor information is considered.

Since the near-wall turbulence is governed by the presence of QSV, it is expected that maximal dimensions of the actuators and largest sensor separation are directly linked to the geometric dimensions of the QSV. Presumably the size of the QSV provides the lower limit for the reduced resolution sensing and actuation.

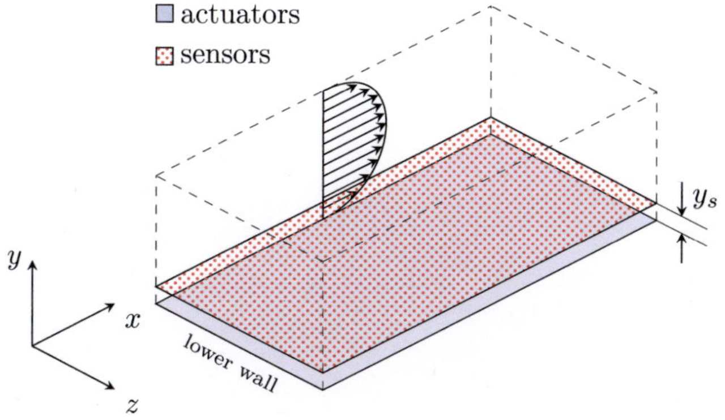

Figure 3.1: Schematic of the classical opposition control.

However, control efficiency can also suffer from other effects linked to the change of the viscous scale. Hence the dependency of the control performance on the resolution properties of the control scheme is to be investigated in the following.

A parametric study with variation of the streamwise and spanwise actuator size is carried out using DNS of a TCF under CFR. For this purpose the actuation at the wall is separated into equidistant discrete regions where the introduced value for control input is kept constant on the entire actuator area (Figure 3.2). A detailed description of the procedure and results are presented in Appendix A.3. It is found that the control performance is barely influenced by the actuator size and reduced sensor resolution for $\Delta x_a^+ < 100$ and $\Delta z_a^+ < 20$, where $R \approx 22-23\%$ with $G = 25-30$ are observed. For longer streamwise actuator extents $100 < \Delta x_a^+ < 400$ a smooth attenuation of the achievable drag reduction rate and a considerable reduction of G is present. A spanwise extent of $20 < \Delta z_a^+ < 40$ shows a more significant deterioration of the control performance: while R is halved to $R \approx 10\%$, the gain is reduced by a factor of 5 in comparison with smaller size actuators. Simulations with longer actuator/sensor extents either fail due to erroneous control input or yield a negative drag reduction rate.

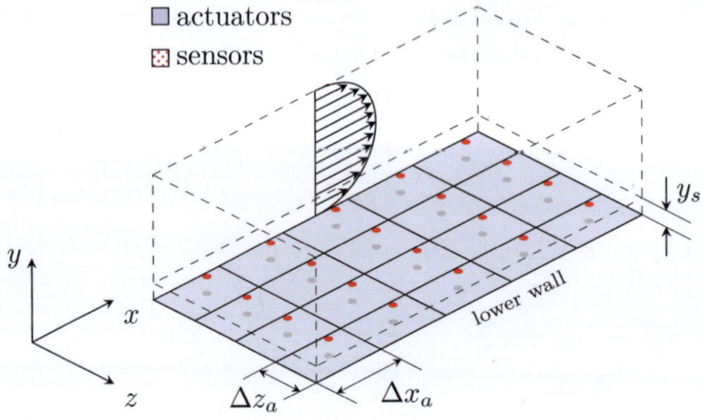

Figure 3.2: Schematic of opposition control scheme with finite size actuators and reduced resolution sensing.

The observed deterioration of the control performance for actuator size exceeding certain dimensions is linked to the control mechanism of the opposition control scheme, namely the suppression of quasi streamwise vortices. QSV feature certain characteristic dimension: a streamwise extent of $\Delta x^+ \approx 200 - 400$ and a spanwise extent of $\Delta z^+ \approx 20 - 50$ [2, 57–59]. Similar dimensions can also be found in the integral lengthscales of the wall-normal velocity component in the near-wall region at $y^+ \approx 10$ [75]. As a consequence, actuators smaller than the average streamwise vortex can affect the structure and eliminate the rotational motion, while actuators larger than the structure can only induce a wall-normal movement of the structures towards or away from the wall. It is found that a thin and long actuator with the streamwise extent approximately 10 times larger than the spanwise extent is preferred for the realisation of the control aimed at the cancellation of the vortical motions or QSV in a turbulent flow. Considering spatial separation of the finite actuators with equidistant gaps of $\Delta x_{a,s}$ and $\Delta z_{a,s}$ in streamwise and spanwise directions as shown in Figure 3.3, the results confirm the previous observation by Fukagata *et al.* [34] that the obtained drag reduction rate depends on the ratio of the active control area to the entire wall area. Summarising the results of the present section it can be concluded that an effective control application demands relatively high coverage of the control area with actuators, while the maximum dimensions and separation of the actuators can be roughly estimated by the geometrical properties of QSV in the near-wall region of the flow.

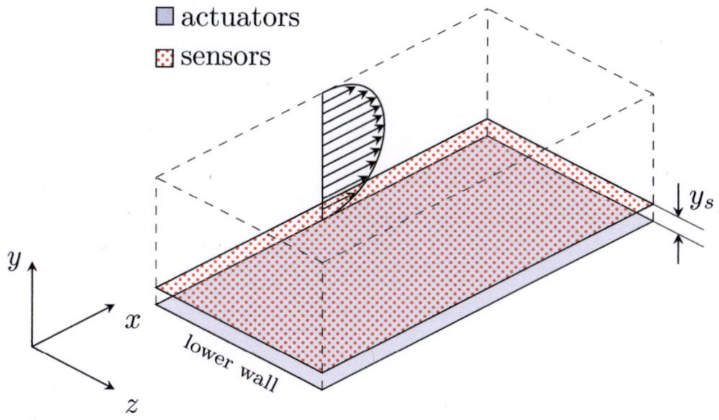

Figure 3.3: Schematic of opposition control scheme with separated finite size actuators and reduced resolution sensing.

3.3 Frequency Limited Sensing

The present section is focused on the investigation of the effects of sensing with limited temporal resolution on the resultant control performance. For this purpose, a systematic change of the sensing interval is performed, while the frequency response of the actuators is assumed to be sufficiently high. The opposition control scheme for wall-normal and spanwise velocity component (equations (2.64) and (2.65)) is utilised as a representative reactive control scheme for the investigation. Control is applied to the entire TCF wall area ($d = 1$) with amplification $\alpha = 1$, while the sensing plane position is fixed to $y_s^+ = 10$. Two domain configurations, 1 and 6 from Table 3.1, at $\mathrm{Re}_\tau = 150$ and $\mathrm{Re}_\tau = 300$, are used for the investigation. The simulations are performed under CFR condition. The computational time step is set to $\Delta t_0^+ = 0.03$ and the total statistical integration time is $\Delta t^+ = 9000$ corresponding to a minimum of 15 eddy turnovers. The maximum sensing frequency in the present numerical simulation is given by the computational time step as $f_0^+ = 1/\Delta t_0^+ = 33.33$. In order to investigate the influence of a limited frequency response of the sensors, we assume that sensing at y_s is not carried out at every computational time step, but only with a certain sensing frequency $f_s = 1/\Delta t_s$, where Δt_s is the time interval between discrete measurements at the sensor location. The local control input during the sensing interval is kept constant and determined based on the last sensor information. Consequently, the opposition control input based on discrete sensing information is defined by

$$u_i(x, 0, z, t) = -\delta_{ij} u_j(x, y_s, z, T), \quad T < t < T + \Delta t_s. \tag{3.2}$$

Here, T is an arbitrary sensing time instant and δ_{ij} denotes Kronecker delta:

$$\delta_{ij} = \begin{cases} 1 & \text{if } i = j, \\ 0 & \text{if } i \neq j. \end{cases} \tag{3.3}$$

This control is referred to as time-discrete opposition control in the present study.

„The resultant drag reduction rate, R, the dimensionless control power input, $P_{in,v}/P_0$, the net energy saving rate, S, and the gain, G, for the wall-normal opposition control operated with the time-discrete sensing are summarised in Figure 3.4. It is found that R decreases slightly down to $f^+ \approx 0.22$, and then drops rapidly, while P_{in}/P_0 shows a weak increase in the region down to $f^+ \approx 0.22$ and then a drastic increase in the region below. Correspondingly, the energy saving rate, S, shows a distribution similar to that of R, while G decreases continuously.

At higher Reynolds number, $\mathrm{Re}_\tau = 300$, both S and G are reduced. The drag reduction rate decreases by about 2% from that at the low Reynolds number for high sensing frequencies. This almost constant deviation holds down to $f^+ \approx 0.33$, below which a rapid decrease of R is observed. Similar behaviour is exhibited for the control power input, $P_{in,v}/P_0$, which also shows a rapid growth for $f^+ < 0.33$. In general, it is concluded that at higher Reynolds number the lowest possible sensing frequency is increased for the time-discrete control scheme and the control performance indices are reduced. This reduction is mainly due to the increase in the pressure fluctuations at higher Reynolds number, which is reflected in increased values of $P_{in,v}$ due to the increase in pressure fluctuations.

The same tests for limited frequency resolution are performed for spanwise opposition control. The corresponding results are shown in Figure 3.5. Spanwise opposition control generally exhibits similar trends to those observed in the wall-normal opposition control. However, it is less sensitive to the reduction of the sensing frequency, so that the reduction down to $f_s^+ \approx 0.08$ is possible without persistent deterioration in R (Figure 3.5). The decrease in S and G for decreased sensing frequencies occurs gradually in contrast to the case of wall-normal control. At higher Reynolds number, the observed trends are similar to wall-normal opposition control, that is, a decrease in S and G is observed. While this decrease in S and G in wall-normal opposition control is mainly due to the increase in pressure fluctuations, these do not influence P_{in} in the spanwise opposition control. In this instance, the observed performance deterioration is caused by Reynolds number dependency of the spanwise velocity fluctuations [53].

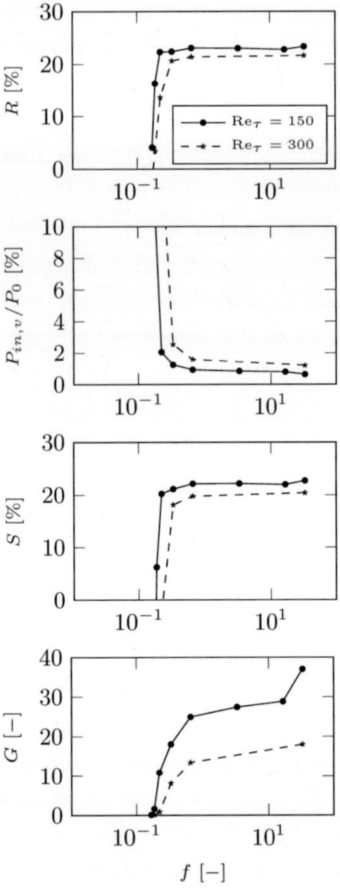

Figure 3.4: Performance indicators for time-discrete wall-normal opposition control with varying frequency resolution of the sensor at $Re_\tau = 150$ and $Re_\tau = 300$ [134, p. 7].

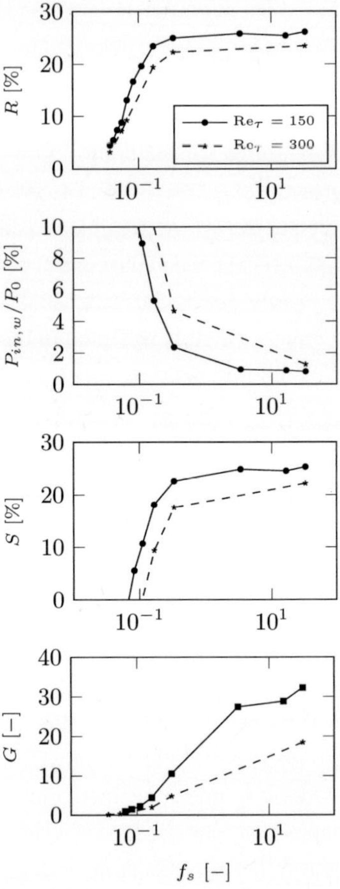

Figure 3.5: Performance indicators for time-discrete spanwise opposition control with varying frequency resolution of the sensor at $Re_\tau = 150$ and $Re_\tau = 300$ [134, p. 7].

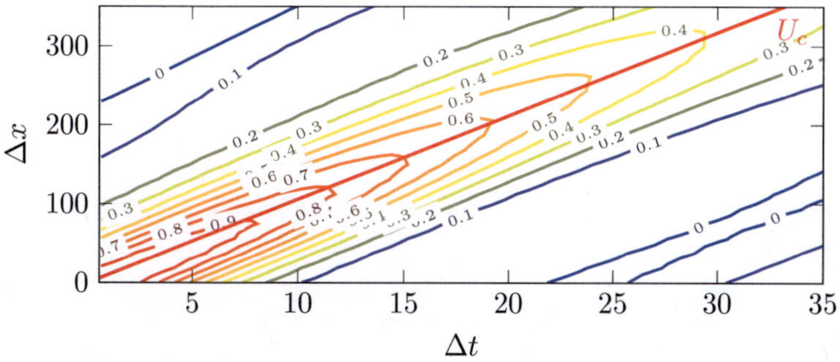

Figure 3.6: Spatio-temporal correlation of the wall-normal velocity component in TCF at the sensing plane position at $\mathrm{Re}_\tau = 150$, $C_{st}\left(v\left(x, y_s, z, t\right), v\left(x+\Delta x, y_s, z, t+\Delta t\right)\right)$ [134, p. 8].

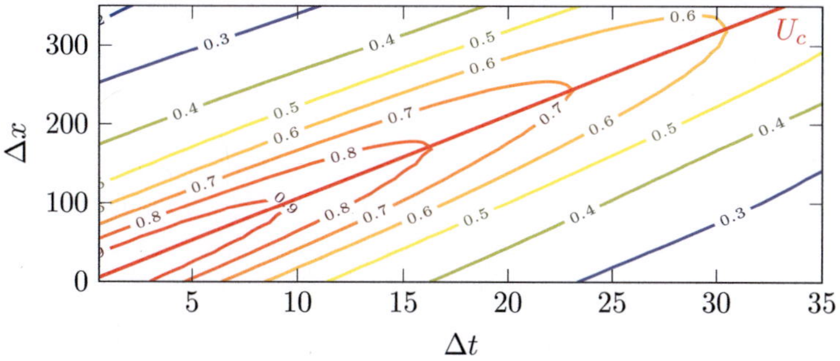

Figure 3.7: Spatio-temporal correlation of the spanwise velocity component in TCF at the sensing plane position at $\mathrm{Re}_\tau = 150$, $C_{st}\left(w\left(x, y_s, z, t\right), w\left(x+\Delta x, y_s, z, t+\Delta t\right)\right)$ [134, p. 8].

The lower limit of the sensing frequency can be explained by the autocorrelation of the sensing quantity. From the spatio-temporal correlations shown in Figure 3.6 and 3.7, it is found that higher correlation values can be sustained for certain Δx and Δt values. Moreover, the maximum values of correlation C_{st} are aligned along a line starting at the origin, where $C_{st} = 1$, with a constant slope estimated as

$$U_c^+ = \frac{\Delta x|_{C_{st,\mathrm{max}}}}{\Delta t^+|_{C_{st,\mathrm{max}}}} \approx 10. \tag{3.4}$$

This quantity coincides with the convection velocity of turbulent structures at the position of the sensing plane, y_s, in the near-wall region [73].

Based on this result, a new approach is followed, where the sensing information is convected downstream with a convection velocity U_c in order to exploit the best sensor signal during the time-discrete operation. As a result, the control input is defined by

$$u_i(x, 0, z, t) = -u_i(x - x_c, y_s, z, T), \quad T < t < T + \Delta t_s, \tag{3.5}$$

where the convection distance is given by $x_c = U_c(t - T)$. This new algorithm is referred to as convected opposition control. In order to realise this control, the sensing information obtained at T is interpolated in the x-direction. In the present study, a simple linear interpolation is employed. It should be emphasised that the control scheme (3.5) assumes continuous actuation (i.e., control input is updated at every computational time step), while the flow information is obtained at every sensing interval Δt_s only.

By applying this method, a decrease in the sensing frequency can be achieved down to $f_s^+ \approx 0.04$, which is more than 800 times below the full frequency sensing, with almost constant energy saving rate as shown in Figure 3.8. The gain for this control scheme remains relatively large for low sensing frequencies. At $\mathrm{Re}_\tau = 300$, the decrease in S and G from those at $\mathrm{Re}_\tau = 150$ is similar to the results observed in the time-discrete control scheme in Figure 3.4. For spanwise opposition control convected control extends the usable sensing frequency region down to $f_s^+ \approx 0.02$ (Figure 3.9).

The limits of applicable sensing frequencies are summarised in Table 3.2. This limit is defined by the sensing frequency at which the input power P_{in} becomes equal to the difference in the pumping power between uncontrolled and controlled flow ($P - P_0$), namely $G = 1$ and $S = 0$. The lower limit of the sensing frequency clearly benefits from the application of the newly introduced convected control scheme.

With respect to the influence of limited sensor frequency response, it can be concluded that the opposition control can still be carried out effectively for sensing frequencies down to $f_s^+ \approx 0.04$, when the convected control scheme is applied. However, this is achieved at the expense of a reduced energy gain compared with the original control scheme. Note that the lowest limit of the sensing frequency can be converted into the length $\Delta x^+ \approx 240$, which corresponds to the longest possible streamwise distance between sensors and actuators as described in Section 3.2. In general, the spanwise opposition control turns out to be slightly less sensitive to the influence of limited frequency resolution than its wall-normal counterpart. This can be attributed to the fact that the spanwise velocity component generally shows larger spatio-temporal correlation than the wall-normal velocity component as shown in Figures 3.6 and 3.7. The difference in correlation might arise from the fact that in the near-wall region the wall-normal velocity fluctuations scale with the distance from the wall due to the wall impermeability, while the tangential velocity components are more influenced by larger structures away from the wall." [134, p. 6–10]

Re_τ	control type	wall-normal control $f_{s,\text{min}}^+$	spanwise control $f_{s,\text{min}}^+$
150	time-discrete	0.1776	0.0603
	convected	0.0340	0.0167
300	time-discrete	0.2240	0.1034
	convected	0.0802	0.0308

Table 3.2: The lower limit of sensing frequency to achieve $G = 1$ for different control types [134, p. 9].

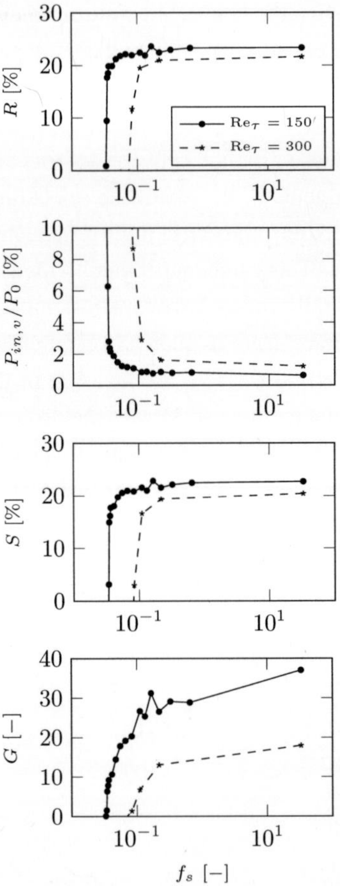

Figure 3.8: Performance indicators for convected wall-normal opposition control with varying frequency resolution of the sensor at $\mathrm{Re}_\tau = 150$ and $\mathrm{Re}_\tau = 300$ [134, p. 9].

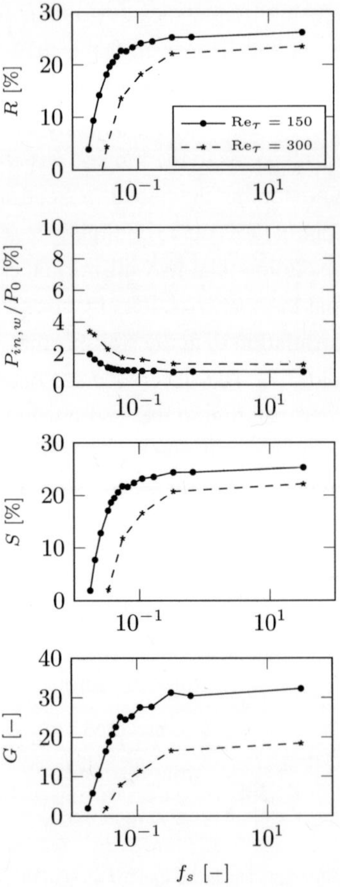

Figure 3.9: Performance indicators for convected spanwise opposition control with varying frequency resolution of the sensor at $\mathrm{Re}_\tau = 150$ and $\mathrm{Re}_\tau = 300$ [134, p. 9].

3.4 Noise Contaminated Sensing

In order to investigate the influence of sensor noise on the control performance, a Gaussian white noise generated by the algorithm developed by Fox *et al.* [32] is superimposed on the sensor signal. Accordingly, the control input given by equations (2.64) and (2.65) is modified as

$$u_i\left(x,0,z,t\right) = -\delta_{ij}u_j\left(x,y_s,z,t\right) + I \cdot n\left(x,z,t\right) \cdot u_{i,rms}\left(y_s\right), \qquad (3.6)$$

where I represents the noise intensity, while $u_{i,rms}\left(y_s\right)$ is the RMS-value of the corresponding velocity component at the sensing plane of y_s in the uncontrolled flow. The random function of $n\left(x,z,t\right)$ prescribes the noise that is spatially and temporally uncorrelated. The probability density function D of $n(x,z,t)$ is given by

$$D(n) = \frac{1}{\sqrt{2\pi}} \cdot e^{-\frac{1}{2}n^2}. \qquad (3.7)$$

For the investigation of the sensor noise influence, a parametric study with variation of the noise intensity, I, and the sensing frequency, f_s, is carried out for time-discrete and convected opposition control. At $\mathrm{Re}_\tau = 150$ in configuration 1 (Table 3.1), the reference noise values, that is $u_{i,rms}\left(y_s\right)$ in equation (3.6), are given by $v^+_{rms}\left(y^+_s = 10\right) = 0.2772$ for wall-normal and $w^+_{rms}\left(y^+_s = 10\right) = 0.7447$ for spanwise control schemes, respectively. Similar to the previous sections, the control is applied to the entire wall area and the simulations are conducted under CFR condition.

„The results of the parametric study for the time-discrete, wall-normal opposition control are shown in Figure 3.10. The drag reduction rate reveals no dependency on the noise intensity. In the contour plot for R we observe a clear lower limit for the sensing frequency, f_s, which can also be seen in the Figure 3.4 and corresponds to $I = 0$ in Figure 3.10. The behaviour of S and G is mainly determined by P_{in}/P_0, as seen in Figure 3.10. At high sensing frequencies, G is significantly reduced even for low noise intensities, due to the increase in pressure fluctuations caused by erroneous suction and blowing. When the maximum sensing frequency is $f^+_s = 33.33$, the wall-normal control yields

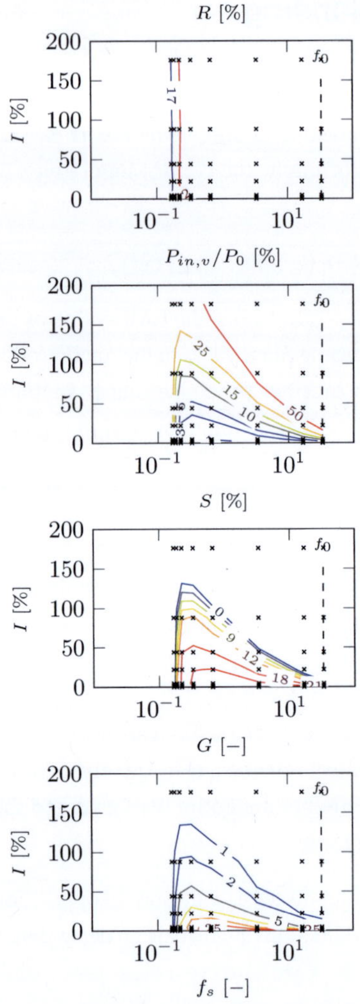

Figure 3.10: Performance indicators (given above each plot) for noise contaminated time-discrete wall-normal opposition control at $\mathrm{Re}_\tau = 150$ [134, p. 10].

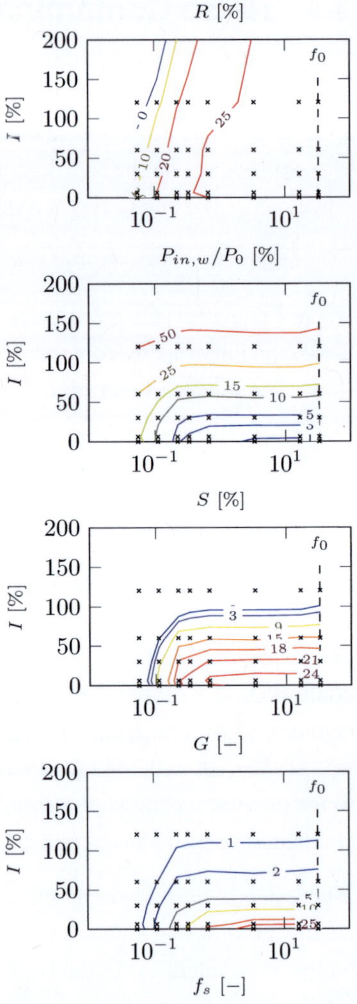

Figure 3.11: Performance indicators (given above each plot) for noise contaminated time-discrete spanwise opposition control at $\mathrm{Re}_\tau = 150$ [134, p. 10].

$G > 1$ only for noise intensities up to $7 - 8\%$. Interestingly, the control scheme becomes less sensitive to noise for lower sensing frequencies, allowing noise intensities up to 30% with $G > 10$ and an energy saving rate of $S \approx 20\%$ for a sensing frequency of $f_s^+ \approx 0.26$. This increased noise resistance of the control scheme at lower sensing frequencies is due to the fact that noisy sensor signals introduce additional pressure fluctuations that enhance the power input, $P_{in,v}$. If additional noise is introduced to the control system at a high frequency, $P_{in,v}$ increases drastically resulting in the reduction of S and G. It should be noted that the drag reduction rate, R, is not influenced by noisy sensor signals provided the sensing frequency is significantly faster than the characteristic time scales of the flow. If noise is applied at lower frequencies, it modifies the perceived flow field such that the effect on the control performance is found for all performance indicators, including R.

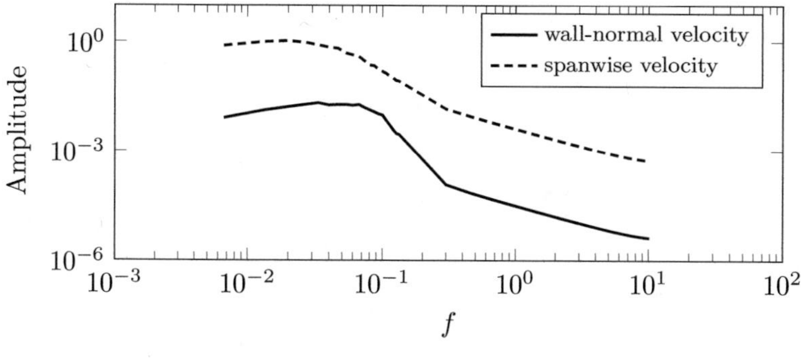

Figure 3.12: Premultiplied energy spectra for velocity components at the sensing plane in an uncontrolled TCF at $\mathrm{Re}_\tau = 150$ [134, p. 11].

The negligibly small noise effect on R at high frequency is explained by the energy spectra of the wall-normal and spanwise velocity fluctuations at the detection plane depicted in Figure 3.12. Since noise is generated for every new sensor update, the noise frequency is defined by the sensor update frequency. The energy spectra of the wall-normal velocity component shows a negligible contribution of frequencies above $f = 0.2$ to the total energy. Hence, the noise effect at higher frequencies is rather weak and does not significantly affect the flow structure or the obtained drag reduction.

When the convected wall-normal opposition control is applied, higher noise levels can be tolerated, as shown in Figure 3.13. This scheme yields energy saving rates above 20% and a gain of 10 – 15 for sensing frequencies down to 0.05 and noise intensities up to 40%. This reduced sensitivity to noisy sensor signals is due to the fact that the actuator input in the convected scheme is determined by interpolation of sensor signals at different streamwise locations. This filters out the random noise, and thus reduces the resultant pressure fluctuations.

The influence of noise contaminated sensor signals on spanwise opposition control is shown in Figures 3.11 and 3.14. In the case of time-discrete sensing (Figure 3.11), the influence of noise on the control performance is almost independent of the sensing frequency in the range of $0.53 < f_s < 33.33$, where positive energy saving rates are found for noise intensities up to 90 – 100%. The energy saving rate and gain are governed by $P_{in,w}/P_0$, while R is hardly influenced by the noise intensity due to the negligible contribution of high-frequency noise to the total energy of the flow. However, in comparison with the wall-normal control scheme the influence of noise in the case of spanwise control is more pronounced, since the peak of the frequency spectrum of the spanwise velocity has a higher amplitude, as shown in Figure 3.12. In contrast to the wall-normal opposition control, the power input for the spanwise opposition control, $P_{in,w}$, is not governed by pressure fluctuations, but by the instantaneous spanwise velocity fluctuation and its wall-normal gradient (see equation (2.61)). The spanwise velocity fluctuations are naturally increased on increasing the noise level, but their dependence on the sensing frequency only becomes apparent for low sensing frequencies, where the erroneous input at the actuator will lead to a deterioration of the drag reduction itself, indicating that the control principle no longer works properly.

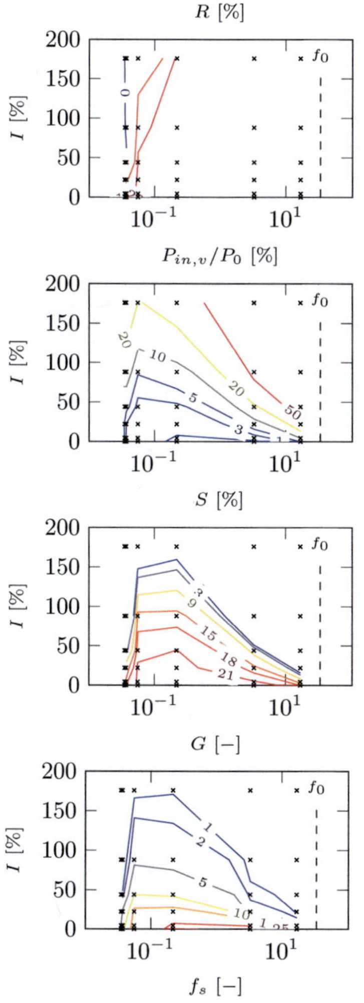

Figure 3.13: Performance indicators (given above each plot) for noise contaminated convected wall-normal opposition control at $Re_\tau = 150$ [134, p. 12].

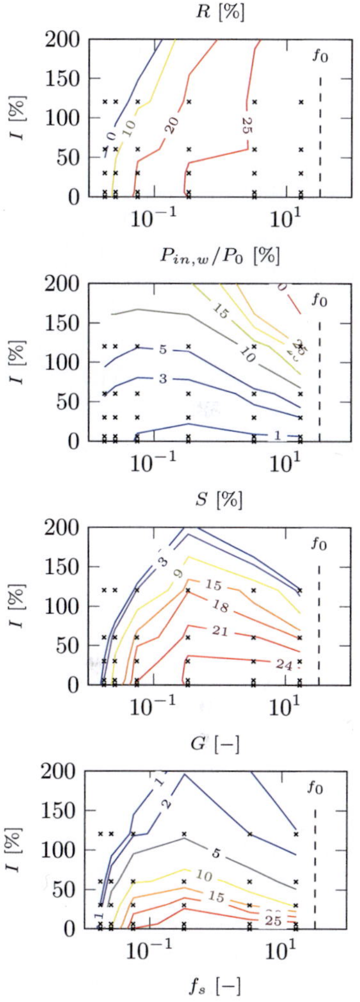

Figure 3.14: Performance indicators (given above each plot) for noise contaminated convected spanwise opposition control at $Re_\tau = 150$ [134, p. 12].

Similar to wall-normal control, the application of the convected control scheme also increases the resistance to noise contaminated sensor signals in the spanwise opposition control (Figure 3.14). The smoothing effect of the interpolation in the convected control scheme, which basically reduces the level of the introduced noise at the actuator, is more pronounced at lower sensing frequencies. The best results for S and G are obtained for sensor frequencies of $0.2 < f_s^+ < 1$, yielding positive values of S and $G > 1$ for noise intensities up to 200%.“ [134, p. 10–12]

3.5 Control with Wall Sensing

Various flow control studies often assume an availability of complete flow information throughout the numerical domain [18, 83]. However, this can obviously not be granted in real applications due to limited simultaneous positioning options for measuring devices and actuators. This also applies to the scheme of opposition control. One of the major limitations of opposition control is the flow measurement located at a certain distance above the wall. Since such measurement inside the flow is extremely difficult in reality, the only reasonable approach is to replace the hypothetical sensors within the flow with sensors at the channel wall [6].

The present section addresses the issue of wall sensing considering three different approaches. In Section 3.5.1 the approach of replacing sensors in the flow field with sensors at the wall based on spatio-temporal correlation is pursued. Section 3.5.2 deals with an attempt to implement wall sensing control utilising the mean impulse response function of the turbulent channel flow [88]. Finally, in Section 3.5.3 the scheme of suboptimal control based on the techniques introduced by Lee *et al.* [83] and Jeon & Choi [56] are investigated and compared with the newly introduced modified schemes.

3.5.1 Correlation Approach

In order to find a suitable wall quantity that can enable an adequate estimation of the velocity information at y_s, an evaluation of a two-point correlation between the sensing plane wall-normal velocity and various wall quantities is considered. It is assumed that if such a quantity correlated to the sensing plane information is found, it can be used instead for control implementation mimicking the opposition control scheme. The following wall quantities can be considered theoretically as measurable: pressure at the wall, p_w, streamwise and spanwise wall shear stress, τ_x and τ_z. Various measurement techniques with high frequency response and small dimensions can be considered suitable for this purpose. The local wall pressure measurement can be implemented using an array of wall-mounted microphones [51], while the wall shear stress measurement in both directions can be conducted using hot film sensors [141] or surface hot wires [136]. A successful application of p_w and τ_z as sensing quantities is demonstrated by Lee *et al.* [82, 83], although these researchers reported that control based on τ_x fails. Jeon & Choi [56] succeeded in the application of a suboptimal control scheme with sensing based on the surface pressure in a framework of a flow over a sphere. The sensing of the spanwise gradient of streamwise wall shear stress, $\partial\tau_x/\partial x$, is also successfully utilised in the work of Frohnapfel *et al.* [33], where reactive feedback control based on the damping of vortical structures using body force is implemented. From the experimental point of view it is known that the measurement of τ_z is rather difficult due to diminishing scales of turbulence towards the wall [138], so a focus on the utilisation of τ_x might be desirable.

The presented approach can be considered similar to the Kalman filtering [65], where a set of temporal measurements is used to predict the current state of an unknown variable. The Kalman filter constitutes a statistical estimator for a linear dynamic system with measurements subject to noise. It is based on the probability theory, theory of stochastic and dynamic systems and can be seen as a learning process, since it is a recursive algorithm [45]. The filter process can be split in two steps - the prediction and the update step.

The prediction step generates an estimate for the variable and then during the update step the estimation is corrected based on the comparison of the measurement with the predicted estimate exploiting weighted average method, where higher weight is assigned to the better estimates. The formal derivation is based on the state evolution equation, which represents the dynamics of the system, and the measurement equation, which describes the noise contaminated measurement model. A detailed overview on the derivation can be found in [46, 63, 65]. The Kalman filter approach can be expanded to a nonlinear version using linearisation of the state equations, also known as the extended Kalman filter [55]. For additional information on the Kalman filter approach in the area of flow control the reader is referred to publications of S. Keshav [70] and Kim & Beweley [72]. The present approach, however, does not consider temporal evolution of the sensor signal and utilises solely statistical information from the computed two-point correlation functions.

According to equation (2.18) the estimation of two-point correlation between the wall-normal velocity at the sensing plane and the considered quantity ϕ at the wall for spatial shifts Δx and Δy is given by

$$C_s\left(\Delta x, \Delta z\right) = \frac{\overline{\phi'\left(x, 0, z, t\right) v'\left(x + \Delta x, y_s, z + \Delta z, t\right)}}{\sqrt{\left(\overline{\phi'\left(x, 0, z, t\right)^2} \cdot \overline{\phi'\left(x + \Delta x, y_s, z + \Delta z, t\right)^2}\right)}}. \tag{3.8}$$

Since control application affects the flow field, the correlation coefficients are computed for uncontrolled flow and flow with an applied wall-normal opposition control scheme. Simulation configuration 5 (Table 3.1) is utilised for computation at $\mathrm{Re}_\tau = 180$ with CFR and the statistics integration time of $\Delta t^+ = 6000$ or 30 eddy turnovers. A configuration with control applied to the entire wall surface ($d = 1$) with amplification $\alpha = 1$ and $y_s^+ = 12$ is utilised for the estimation.

Figure 3.15 shows the correlation distribution between $v(y_s)$ and τ_x. For the uncontrolled flow the quantities are shown to be maximally correlated for $\Delta x^+ = 50$ and $\Delta z^+ = 0$. These shifts provide a correlation coefficient above 0.5 and can be considered to provide a good estimation of the near-wall dynamics [26, 33]. The negative sign of the correlation is explained by the fact that a positive wall-normal motion or ejection leads to the transport of low-speed fluid away from the wall region and thus the wall shear stress is reduced. In contrast, a negative $v(y_s)$ or a sweep corresponds to an increase in the streamwise velocity gradient at the wall. The optimal shift $\Delta x^+ = 50$ is most likely related to the time lag between the change of the τ_x due to the presence of a QSV and the occurrence of an event (ejection or sweep) at the sensing plane position. This time lag translates to the length of $\Delta x^+ = 50$, since a streamwise convection of the flow occurs [73, 111]. A correlation coefficient of 0 is found for $\Delta z^+ \approx \pm 20$, which corresponds to the radius of the QSV. A significant change in the correlation distribution is observed for the flow field with active opposition control. The correlation sign is globally inverted and the highest correlation exceeds 0.8 showing that the wall information obtained is entirely dominated by the imposed control. The positive correlation is linked to the fact that opposition control enforces $v(y = 0) = -v(y_s)$, so a positive $v(y_s)$ means negative $v(y = 0)$ leading to a lower streamwise velocity gradient and reduced wall shear stress. The optimal position is shifted upstream to $\Delta x^+ = 20$, showing a shorter time lag corresponding to a shorter distance due to the prominent control effect. Similar to the uncontrolled flow, a correlation of 0 can be found for the distance of $\Delta z^+ \approx \pm 20$, again corresponding to the radius of the typical QSV.

Figure 3.16 depicts the two-point correlation distribution between $\partial \tau_x / \partial z$ at the wall surface and $v(y_s)$. A distribution similar to dimensional composition and magnitudes observed for correlation between $v(y_s)$ and τ_x is evident. For the uncontrolled case, positive correlation is present for negative Δz values and vice versa. This is linked to the rotational nature of the QSV, where positive $v(y_s)$ corresponds to a negative $\partial \tau_x / \partial z$ for negative Δz and with reversed order for positive Δz. The controlled case also changes the correlation sign since opposition control introduces structures similar to QSV but in the opposite direction at the wall area. The maximal values of above 0.4 and 0.6 for uncontrolled and controlled cases, as well as the distribution in the streamwise direction with highest values around $\Delta x^+ \approx 50$ and $\Delta z^+ = \pm 15$, are in a good agreement with the correlation distribution for the same quantity at $y_s^+ = 15$ presented by Frohnapfel *et al.* [33].

The two-point correlation between spanwise wall shear stress at the wall and velocity $v(y_s)$ is presented in Figure 3.17. The distributions resemble those in Figure 3.16, where again the rotational behaviour of the QSV can be observed. Considering vortical structures, positive τ_z corresponds to positive $v(y_s)$ for positive Δz, so a positive correlation is expected for a positive Δz and vice versa. The maximum correlation value of above 0.5 can be found for a small negative Δx and $\Delta z^+ \approx \pm 15$. Interestingly, the relatively high maximal correlation does not change the sign and even slightly increases (above 0.6) for the controlled case. It seems that opposition control does not significantly affect the spanwise wall shear stress.

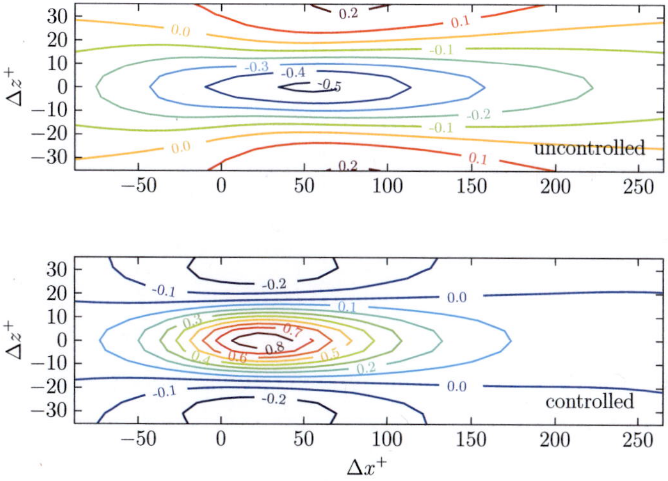

Figure 3.15: Two-point correlation between $\tau_x(y = 0)$ and $v(y_s^+ = 12)$.

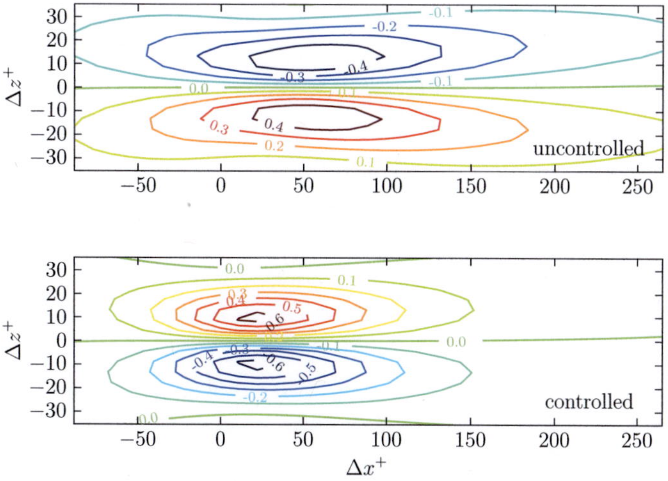

Figure 3.16: Two-point correlation between $\frac{\partial \tau_x}{\partial z}(y = 0)$ and $v(y_s^+ = 12)$.

Finally, the two-point correlation between wall pressure and $v(y_s)$ is presented in Figure 3.18. In the uncontrolled case the maximum correlation reaches a value of about 0.3, suggesting that the prediction based on the pressure in the uncontrolled flow is rather difficult. Presumably this happens due to high sensitivity of pressure on the velocity fluctuations present in the flow field. Negative correlation is observed for $\Delta x^+ = -40$ confirming an intuitive assumption that negative velocity at the sensing plane introduces a positive pressure at the wall surface. A complete change of the correlation distribution is observed for the controlled case with higher maximum values of approximately 0.6.

Since the spanwise wall shear stress, τ_z, is the only quantity providing similarly a high correlation that is not changing its sign during the control application, an attempt is made to replace the sensing plane information with $\tau_z(y = 0)$. Similarly to the definition in equation (2.64) the control is the defined by

$$v\,(x, y = 0, z, t) = -\alpha\tau_z\,(x + \Delta x, y = 0, z + \Delta z, t)\,, \tag{3.9}$$

where Δx and Δz correspond to the streamwise and spanwise offset of the sensor relative to the actuator. The values are chosen at $\Delta x^+ = -11$ and $\Delta z^+ = 18$ based on the location of the maximal correlation values found in Figure 3.17. The best working configuration for the proposed scheme is found at $\alpha = 0.07$; control-loop shows strong instabilities for higher values. The control yields maximum of $R \approx 2\%$, $S \approx 1.9\%$ and $G \approx 25$. High energy gain shows that the amplitude of the applied control is extremely low, which explains poor control performance. Supposedly, in spite of the fact of high observed correlation values, τ_z cannot provide sufficiently accurate information about the velocity state at the sensing plane position. Since τ_z is very small in the wall vicinity, the introduction of the wall-normal velocity component at the wall substantially affects the distribution of τ_z, so the two-point correlation that can be measured in the controlled case no longer reflects the relationship between τ_z and $v(y_s)$, but rather shows the dependency of the sensor signal on the control input.

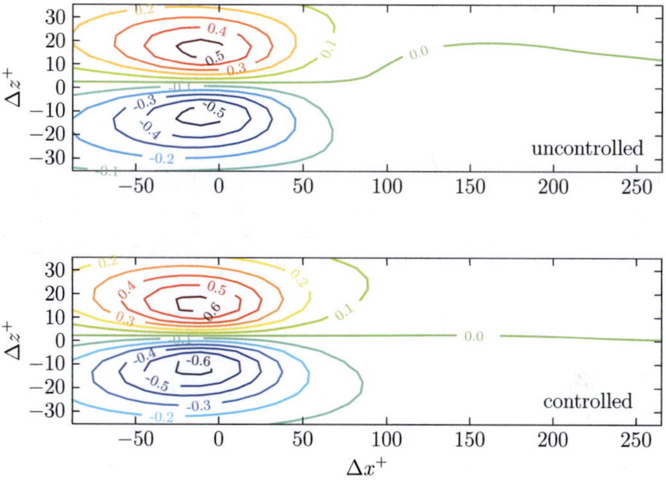

Figure 3.17: Two-point correlations between $\tau_z(y = 0)$ and $v(y^+ = 12)$.

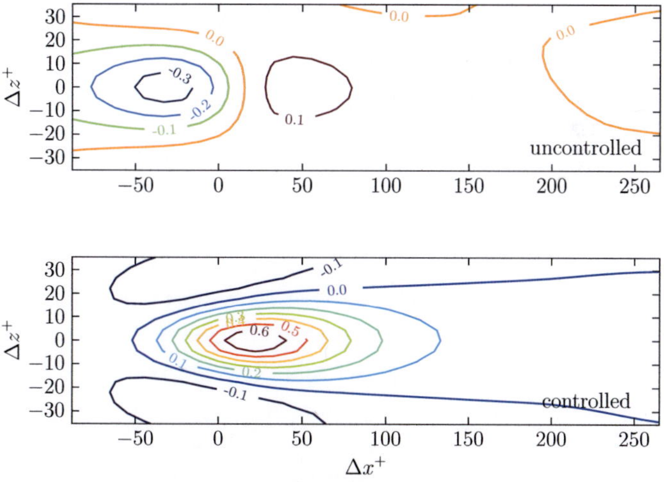

Figure 3.18: Two-point correlations between $p(y = 0)$ and $v(y^+ = 12)$.

As a next step, a control input based on the streamwise shear stress, namely the spanwise gradient of streamwise wall shear stress, $\partial\tau_x/\partial z$, is considered. Since it is known that the correlation distribution changes the sign for the flow with applied opposition control (Figure 3.16), $\partial\tau_x/\partial z$ cannot be used for a prediction of the wall-normal velocity in a straightforward manner. An attempt is made to overcome this limitation using a configuration where sensors on the wall surface are spatially separated from the actuators, as shown in Figure 3.19. Such an application of spatial separation aims at the implementation of a more realistic control technique and enables a reduction of pollution of the sensor information due to the active control.

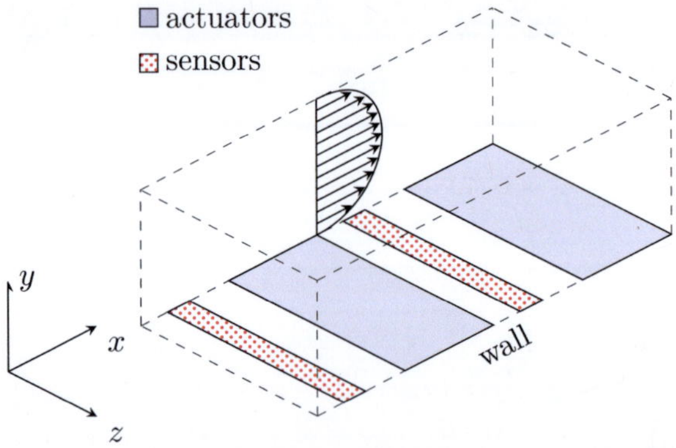

Figure 3.19: Schematic of control based on wall information.

In order to investigate the effects of different sensor and actuator arrangements, four characteristic control configurations are introduced. These configurations assume varying arrangements of actuators and sensors in the streamwise direction, while continuous constant distribution of sensors and actuators is considered in the z-direction. The following scenarios are considered in order to approach the desired configuration, where only wall sensing is used:

- *Partial control*
  original opposition control is applied on a certain part of the total wall area only

- *Reduction of spatial sensing resolution*
  the number of sensors is reduced while control is applied at the entire wall

- *Partial control with reduced spatial sensing resolution*
  combination of the first two steps, such that the sensor location differs from the actuator location

- *Control input based on wall information*
  measurements in the flow field are replaced by sensor information from the wall

A detailed parametric study on the three intermediate configurations with control input based on sensing plane information and the fourth configuration with wall sensing is presented in Appendix A.4. Table 3.3, where the corresponding results for $\Delta x_a^+ = 147$ at $\mathrm{Re}_\tau = 150$ are presented, summarises the influence of the arrangement variation on the control performance of the schemes investigated. Obviously, a significant deterioration of control performance occurs when a reduction of the active control area (partial control) is introduced (as discussed in Section 3.2), while a reduction of the sensing resolution only slightly affects the control performance. In spite of these limitations, control schemes with sensing plane information still yield a significant drag reduction rate and net energy gain. However, utilisation of wall information for the estimation of control input shows a major impact on the control performance, which is remarkably reduced to $R = 5.3\%$ for this case. Even though the sensors are placed inside the relaxation sections, it seems that control input estimated with wall sensing cannot entirely reproduce the flow field information available at the sensing plane.

control type	control input	R [%]	S [%]	G
partial control with full sensing resolution	sensing plane	14.2	13.6	23.1
reduction of spatial sensing resolution	sensing plane	22.1	21.3	27.4
partial control with lower sensing resolution	sensing plane	16.4	15.3	20.8
partial control with lower sensing resolution	wall sensing	5.3	5.1	20.6

Table 3.3: Comparison of control performance for variation of the spatial arrangement and control input estimation. Best values for $\Delta x_a^+ = 147$ at $\mathrm{Re}_\tau = 150$ are presented.

3.5.2 Impulse Response Approach

In the previous section the utilisation of simultaneous wall sensing and actuation proved to be a great challenge. Control application drastically pollutes the wall sensor signal, reducing the accuracy of the prediction. It is shown that an introduction of a certain distance between sensors and actuators reduces the influence and provides acceptable control performance. The present section deals with another possibility for the reduction of the control influence on the acquired sensor information, namely the filtering of the sensor signal using the impulse response function (IRF).

The approach is based on the investigation of Luchini *et al.* [88], where a linear response of a TCF to flow perturbations is determined using DNS. The study considers an introduction of a small disturbance, for example, in the form of a wall-normal velocity at the wall and measures the linear flow response throughout the simulation domain. Since the IRF provides information about the influence of the disturbance on the wall quantities, it could also be used to estimate the influence of the actuation on the sensor signal. The investigation of Luchini *et al.* [88] consider a mean impulse response function of the flow to the imposed wall-normal velocity:

$$v(x, y = 0, z, t) = \epsilon\delta_\epsilon(x)\,\delta_\epsilon(z)\,\delta_\epsilon(t), \tag{3.10}$$

where δ_ϵ denotes the Dirac delta function and ϵ is the amplitude of the impulse, which is chosen to be small enough for the estimation of the linear response. The resultant IRF is a tensor $H_{ij}(x, y, z, t)$, which describes an average effect of the velocity u_j applied at the wall surface on the velocity u_i at an arbitrary position in the simulation domain. The estimated IRF for τ_x at $\mathrm{Re}_\tau = 180$ with a spatio-temporal range of $0 \leq t^+ \leq 32$, $0 \leq x^+ \leq 237$ and $-8 \leq z^+ \leq 8$ has kindly been provided by Prof. Dr. M. Quadrio for research purposes. The IRF for τ_x is given as

$$H_x(x, z, t) = H_{12}(x, y = 0, z, t), \tag{3.11}$$

so the mean response $\tilde{\tau}_x$ due to the imposed wall-normal velocity at the wall is computed using a convolution integral

$$\tilde{\tau}_x(x, z, t) = \int_0^{t'}\int_0^{z'}\int_0^{x'} H_x(x', z', t')\, v(x - x', y = 0, z - z', t - t')\,\mathrm{d}x'\mathrm{d}z'\mathrm{d}t'. \tag{3.12}$$

Assuming a linear response of the flow to the applied control, the filtered wall shear stress, $\hat{\tau}_x$, is estimated by a simple subtraction

$$\hat{\tau}_x = \tau_x - \tilde{\tau}_x. \tag{3.13}$$

Based on the filtered information the control scheme with utilisation of τ_x

$$v(x, y = 0, z, t) = -\alpha\hat{\tau}_x(x - \Delta x, y = 0, z - \Delta z) \tag{3.14}$$

and of $\partial\tau_x/\partial z$

$$v(x, y = 0, z, t) = -\alpha\frac{\partial\hat{\tau}_x}{\partial z}(x - \Delta x, y = 0, z - \Delta z) \tag{3.15}$$

are defined. The spatial shifts are chosen based on the two-point correlations from Figures 3.15 and 3.16 with $\Delta x^+ = 55$ for both schemes. The spanwise shift is fixed at $\Delta z^+ = 0$ and $\Delta z^+ = 18$ for τ_x and $\partial\tau_x/\partial z$ sensing, respectively.

An application of both control schemes in configuration 3 (Table 3.1) at $Re_\tau = 180$ on the entire wall surface fails leading to numerical instabilities during the control activation at all tested amplitudes α. The issue is probably linked to the strong assumption of a linear response, which cannot be entirely expected within the framework of blowing and suction applied at the wall surface. Therefore, the approach of IRF-filtering is tested in a configuration of partial control with wall sensing similar to the last configuration used in previous section (Figure 3.19). Two cases are investigated in a CFR TCF at $Re_\tau = 150$ in configuration 1 (Table 3.1). The schematic and distance parameters are presented in Figure 3.20.

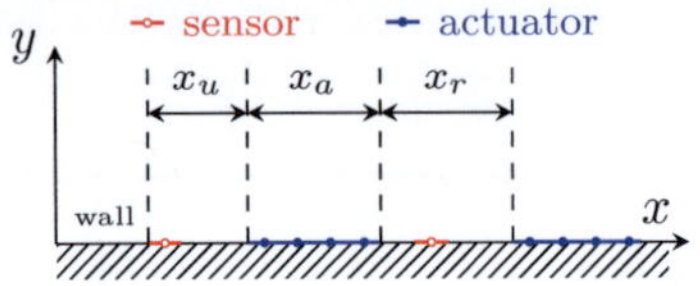

Case	Δx_r^+	Δx_a^+	Δx_u^+
1	88	88	33
2	55	121	33

Figure 3.20: Cases considered for IRF-filtered control with wall-sensing.

In the first case a long recovery section ($\Delta x_r^+ = 88$) is utilised in order to ensure that sensor information at the wall is not polluted by control actuation. The length is chosen based on the results presented in Section 3.5.1, where the minimal applicable length of the recovery region is reported to be $\Delta x_r^+ = 74$. The size of the actuation region is chosen to be the same length, $\Delta x_a^+ = \Delta x_r^+ = 88$, so the control is applied to 50% of the wall area. The sensors are placed upstream of actuators with $\Delta x_u^+ = 33$.

The second case utilises a shorter recovery section of $\Delta x_r^+ = 55$, while the elongation of the actuation region is $\Delta x_a^+ = 121$. The sensor position is kept at the same upstream distance of $\Delta x_u^+ = 33$. It is expected that wall sensing exhibits significantly poorer prediction ability in this case due to the shorter recovery region. At the same time, however, the actuation coverage is 70% of the wall area, which may naturally lead to higher drag reduction rates.

Both control cases are tested with the convected wall information control scheme (similar to definition (3.5)) determined for τ_x-sensing as

$$v(x, y = 0, z, t) = -\alpha\tau_x(x - \Delta x, z, t - \Delta t), \qquad (3.16)$$

and for $\partial\tau_x/\partial z$-sensing as

$$v(x, y = 0, z, t) = -\alpha\frac{\partial\tau_x}{\partial z}(x - \Delta x, z - \Delta z, t - \Delta t), \qquad (3.17)$$

with the time delay estimated from the convection velocity $U_c^+ = 10$:

$$\Delta t^+ = \begin{cases} 0 & \text{for } \Delta x^+ \leq 55, \\ \frac{\Delta x^+ - 55}{U_c^+} & \text{for } \Delta x^+ > 0. \end{cases} \qquad (3.18)$$

The control scheme definition and shifts for sensor positioning are chosen based on the two-point correlations from Figures 3.15 and 3.16, where an optimal position $\Delta x^+ = 55$ for both sensing quantities, while $\Delta z^+ = 0$ for τ_x-sensing and $\Delta z^+ = 18$ for $\partial\tau_x/\partial z$-sensing yield the highest correlation coefficients.

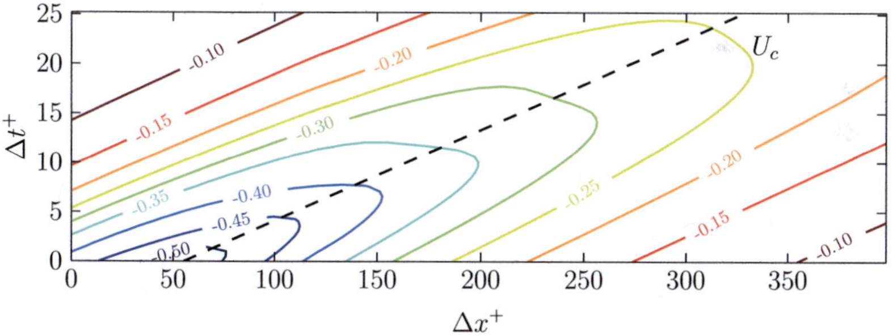

Figure 3.21: Spatio-temporal correlation between τ_x at the wall and $v(y_s^+ = 12)$.

Figure 3.21 depicts the spatio-temporal correlation between τ_x at the wall and wall-normal velocity at the sensing plane $y_s^+ = 12$ for an uncontrolled TCF at $\text{Re}_\tau = 150$. The distribution shows that convection velocity can be used in order to improve the sensor information validity for wall sensing of τ_x. A similar distribution of the spatio-temporal correlation between $\partial\tau_x/\partial z$ at the wall and $v\left(y_s^+ = 12\right)$ is also present when the spanwise shift $\Delta z^+ = 18$ from Figure 3.16 is taken into account.

The variation of control amplitude, α, for τ_x-sensing reveals that the best performance is achieved at $\alpha = -0.15$ for the case with a long recovery area (case 1). For this amplification $R = 4.1\%$, $S = 4\%$ and $G = 34$ are observed. Higher control amplitudes lead to numerical instabilities and a breakdown of the simulation. Simulations with shorter recovery sections (case 2) fail. Due to the poor control performance of the control scheme based on τ_x-sensing, the investigation continues with $\partial\tau_x/\partial z$-sensing only.

Case	IRF-filtering	$R[\%]$	$S[\%]$	G
1	off	9.4	8.3	8.5
	on	9.3	8.0	7.2
2	off	−1.2	−4.2	−0.4
	on	10.0	8.9	9.0

Table 3.4: Control performance for convected scheme with $\partial\tau_x/\partial z$-sensing at the wall.

A control scheme based on $\partial\tau_x/\partial z$-sensing is shown to be more stable operating with $\alpha = -12$ and yielding $R = 9.4\%$, $S = 8.3\%$ and $G = 8.5$ for the long recovery area (case 1). A shorter recovery area (case 2) exhibits a drag increase with $R = -1.2\%$, which is again related to the strong pollution of the sensor information. Filtering using IRF (equation (3.13)) is applied to both cases with $\partial\tau_x/\partial z$-sensing resulting in a significant improvement in the short recovery section case, while the control performance in the case of longer recovery section remains unchanged. A comparison of simulation results is presented in Table 3.4. The unresponsiveness of the long recovery section case is linked to the weak or even non-existent pollution of the sensing information and therefore does not benefit from the filtering. In the short recovery section case the control performance can be drastically influenced by the application of IRF-filtering leading to $R = 10\%$, $S = 8.9\%$ and $G = 9$. The IRF-filtered case 2 slightly outperforms case 1 due to the larger control area coverage. However, this slight improvement from $R = 9.4\%$ for the long recovery section case without IRF-filtering to $R = 10\%$ for short recovery section with IRF-filtering cannot be considered reasonable since an additional signal processing step is required in the latter case. An introduction of the IRF-filtering in a realistic control application would translate into a much longer time lag of the control loop or the necessity for a faster signal processing system. Thus, in spite of the successful application of the IRF-filtering, the simpler configuration with a longer recovery region and absent filtering might be preferred.

3.5.3 Suboptimal Control with Wall Sensing

As discussed earlier, a suboptimal control scheme represents another possibility for wall information utilisation for the estimation of control input. For this scheme the control input is derived analytically or estimated numerically based on the minimisation of a predefined cost functional. In the following a suboptimal control scheme is compared with the control schemes with wall-sensing based on the correlation approach (Section 3.5.1) or exploiting IRF-filtering (Section 3.5.2). Two suboptimal control definitions are examined in the present investigation: a classical suboptimal control scheme by Lee *et al.* [83] based on the derivation of the control input in spectral space where pressure and spanwise wall-shear stress are used as a sensing quantity and the more recent suboptimal control scheme by Jeon & Choi [56] with derivation based on the response of the flow domain with pressure providing the sensor information.

In general, the definition of the control input for suboptimal control is given by the convolution integral

$$v\,(x, y = 0, z, t) = \alpha \int_0^{x'} \int_0^{z'} W\,(x', z') \cdot \varphi\,(x + x', y = 0, z + z', t)\,\mathrm{d}x'\mathrm{d}z', \tag{3.19}$$

where φ represents the sensing quantity and W defines the weight function. The equation can be also rewritten in discrete form and physical space for the control input at the wall as

$$v\left(x_i, z_k\right) = \alpha \sum_{i'} \sum_{k'} W_{i'k'} \cdot \varphi\left(x_{i+i'}, z_{k+k'}\right), \tag{3.20}$$

where i and k represent the discretising indices in the streamwise and spanwise directions, respectively. The difference between the two considered suboptimal control schemes is the variation of the weight function, W, obtained from different derivation procedures. In order to compare the suboptimal control scheme with schemes from previous sections that utilise various wall shear stress related quantities, suboptimal control based on the spanwise wall shear stress is considered in the following.

Firstly, the suboptimal control scheme presented by Lee *et al.* [83] is tested. The cost functional is specified in a way that an increase of the pressure gradient in the spanwise direction of the near-wall region is desired. The definition is based on the observation of a similar effect when opposition control is employed in a TCF. Applying the Fréchet differential [29] to the cost functional those researchers derived the necessary control input

$$v(y=0) = u_i(y=0) = \varphi\left(x, 0, z, t\right)\delta_{i2} \quad \text{for} \quad i = 2, \tag{3.21}$$

with sensing of the spanwise wall shear stress

$$\hat{\varphi} = \alpha \frac{ik_z}{k} \cdot \widehat{\frac{\partial w}{\partial y}}\left(y=0\right), \tag{3.22}$$

where "^" denotes the Fourier coefficient representation and k_x and k_z stand for the streamwise and spanwise wavenumbers with

$$k = \sqrt{\left(k_x^2 + k_z^2\right)}. \tag{3.23}$$

The spectral space representation can be numerically transformed into the discrete convolution integrals in the form of equation (3.20) with an appropriate weight function $W_{i'j'}$ for spanwise wall shear stress control. The paper [83] provides the weight function distribution, which is truncated to $(i \times k) = (3 \times 6)$ values with grid spacings $\Delta x^+ = 40$ and $\Delta z^+ = 13$.

Secondly, the derivation based on control proposed by Jeon & Choi [56] is applied to the configuration of a TCF with $\partial w/\partial y$ utilised as sensor information. The cost functional is defined as

$$J\,(v(y=0)) = \int_x \int_z \left(\frac{\partial w}{\partial y}\,(y=0) \right)^2 \mathrm{d}x\mathrm{d}z, \tag{3.24}$$

so by employing a gradient algorithm for the Fréchet differential of the cost functional followed by derivation of the control input based on discretised Navier–Stokes equations the definition (3.20) can be derived. In this case, however, the weight function, W, is the solution to the following equations:

$$\zeta_i + \frac{\Delta t_c}{2}\frac{\partial W}{\partial x_j} - \frac{\Delta t_c}{2\mathrm{Re}}\frac{\partial^2 \zeta_i}{\partial x_j \partial x_j} = 0, \tag{3.25}$$

$$\frac{\partial \zeta_i}{\partial x_i} = 0, \tag{3.26}$$

with boundary conditions

$$\begin{cases} \zeta_i = \delta_{i2} & \text{for } y = 0, \\ \zeta_i = 0 & \text{elsewhere.} \end{cases} \tag{3.27}$$

Thus Δt_c represents the control update time interval. For detailed a mathematical derivation the reader is referred to the corresponding publications [17, 56, 93]. The calculation can be performed numerically and has to be carried out only once, since the solution is not time-dependent.

Both suboptimal control schemes are implemented and tested for configuration 4 (Table 3.1) under CFR condition at $\mathrm{Re}_\tau = 180$. The time step is set to $\Delta t_0^+ = 0.018$ and statistical averaging is performed during $\Delta t^+ = 14000$ corresponding to 38 eddy turn overs. The control is applied to the entire wall area. The amplification α is adjusted whenever the control input is updated, in order to maintain a prescribed value v^+_{rms} on the entire wall area. Figure 3.22 shows the comparison of the control performance achieved by the two suboptimal control schemes and the opposition control scheme. Suboptimal control based on the weight function provided by Lee *et al.* [83] yields approximately 5% less R and S than the opposition control scheme with a minor deterioration of G. The control scheme based on the numerical estimation of W proposed by Jeon & Choi [56] performs slightly worse, but still exhibits comparable performance indices. It has to be mentioned that the presented comparison (Figure 3.22) shows the most successful cases with identical control parameters $\Delta t_c^+ = 0.36$ and $v_{\mathrm{rms}}^+ = 0.1375$.

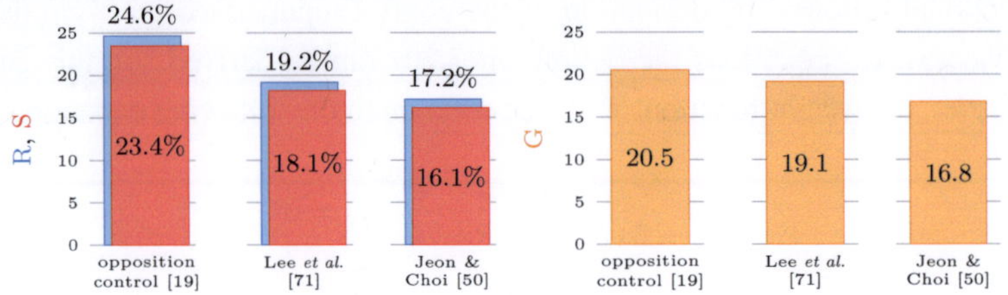

Figure 3.22: Control performance of opposition control scheme and suboptimal control schemes with wall sensing of $\partial w/\partial y$.

It is found that the following three control parameters strongly influence the achievable control performance:

- control update time interval Δt_c
- control amplitude α
- truncation of the weight function W

Figure 3.23 presents the influence of the chosen Δt_c on the achievable drag reduction rate. Poor control performance with $R \approx 7\%$, negative S and $G < 1$ is observed for the case where the control time interval is similar to the simulation time step. An increase in Δt_c leads to a significant improvement of the control performance. The region $10\Delta t_0 < \Delta t_c < 50\Delta t_0$ is found to provide satisfactory results that can be compared with the performance achieved by the opposition control scheme. It has to be emphasised that in this case the control update time interval is still smaller than one viscous time unit ($0.2 < \Delta t_c^+ < 1$). For longer update intervals ($\Delta t_c > 50\Delta t_0$) a deterioration of the obtainable drag reduction rate is observed. The last trend is similar to the behaviour observed for the opposition control scheme with frequency limited sensing (Section 3.3). However, the opposition control scheme does not demonstrate a decay of the control performance for $\Delta t_c \approx \Delta t_0$. It is most likely that this behaviour is linked to the occurrence of strong pressure fluctuations in the near wall region, which are induced by the frequently changing control actuation. As a consequence, strong pollution of the wall information is introduced, affecting the control loop performance. The information provided by sensors does not give an appropriate estimation of the flow state, resulting in a very high control input power, extremely low G and even negative S values for Δt_c close to Δt_0.

Different prescribed values of v_{rms}^+ are tested with the considered suboptimal control schemes. The influence of v_{rms}^+ variation on the drag reduction rate for fixed $\Delta t_c^+ = 0.36$ is shown in Figure 3.24. Both suboptimal control schemes provide the highest performance in the range $0.1 < v_{\mathrm{rms}}^+ < 0.15$, which corresponds to the v_{rms}^+ of the control input provided by the opposition control scheme with the sensing plane at the position $10 < y_s^+ < 15$. The enforced control amplitude obviously affects the achieved G, since low amplitude directly translates into low gain values. Correspondingly, a monotonically decreasing G is observed for increasing v_{rms}^+ values.

$(i \times k)$	(64×64)	(4×14)	(2×11)	(2×8)	(1×14)	(1×8)
$R[\%]$	16.4	15.7	12.7	13	8.7	8.1
$S[\%]$	15.4	14.6	11.5	11.8	7.4	6.7
G	15.2	14.4	11	10.6	6.3	5.8

Table 3.5: Influence of the weight function truncation on the achievable control performance indices with $\Delta t_c^+ = 0.36$, $v_{\mathrm{rms}}^+ = 0.1375$ and grid spacings $\Delta x^+ = 9.2$, $\Delta z^+ = 3.7$.

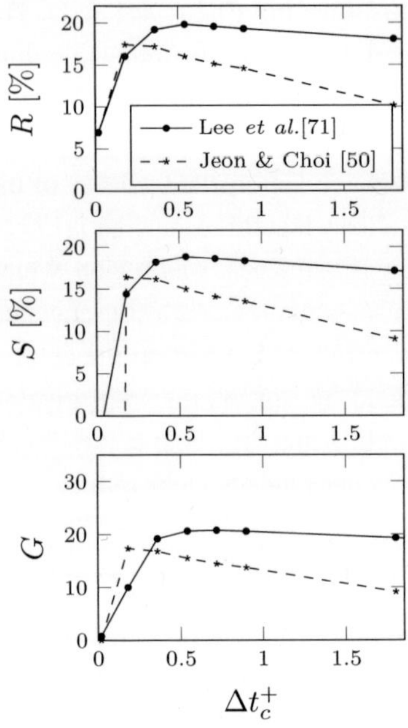

Figure 3.23: Influence of the control update time interval on the achievable control performance with control amplitude fixed at $v_{\mathrm{rms}}^{+} = 0.1375$.

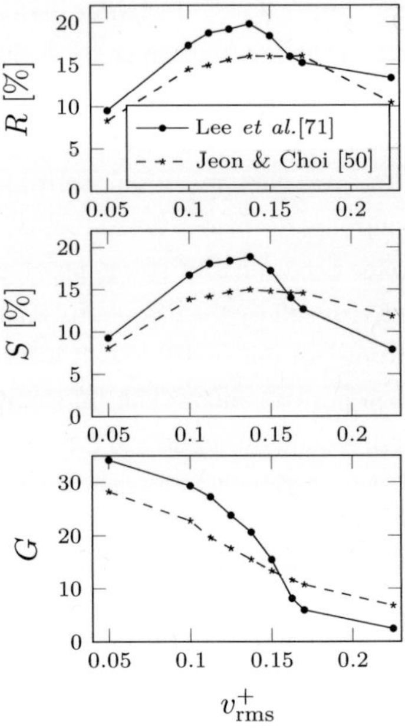

Figure 3.24: Influence of the control amplitude on the achievable control performance with control update time interval fixed at $\Delta t_c^{+} = 0.36$.

The weight function obtained with numerical estimation according to equation (3.26) is symmetrically distributed relative to the point of origin. The size of the function is defined by the size of the simulation domain or the size of the horizontal cross section, $(N_x \times N_z)$. Due to the symmetry of the distribution a sufficient representation of the weight function constitutes the size of $(i \times k) = (N_x/2 \times N_z/2)$, corresponding to a matrix with (64×64) elements in the considered simulation domain. Since the calculation of the convolution integral is computationally expensive, a truncation of the weight function is considered. Based on the fast decay of the function values for increasing x' and z', Lee *et al.* [83] reported that such truncation is reasonable and provides a matrix with only (3×6) elements. Table 3.5 shows the influence of the weight matrix truncation on the resulting control performance for the suboptimal control scheme based on the derivation by Jeon & Choi [56]. It is found that the weight function can be cropped to (4×14) elements with a minor reduction of the achievable control performance. Further deterioration is observed for smaller matrices with a significant decrease of the control performance found only for $N_x < 2$.

	correlation based control	correlation based control with IRF-filtering	suboptimal control
sensing	$\partial\tau_x/\partial z$	$\partial\tau_x/\partial z$	τ_z
control area [%]	50	70	100
R[%]	9.4	10	19.2
S[%]	8.3	8.9	18.1
G	9.5	9.0	16.8

Table 3.6: Comparison of the investigated control schemes with wall information sensing.

Table 3.6 presents comparison of the best obtained control performance for three different control schemes with wall sensing: control based on a correlation approach with convection of sensor information, the same scheme with applied IRF-filtering and suboptimal control by Lee *et al.* [83]. Since the control area varies for these cases, a direct comparison is rather difficult. However, assuming that the ratio of control area to entire wall area is directly proportional to the achievable drag reduction rate [34], one can conclude that all three cases provide very similar control performance.

3.6 Concluding Remarks

Chapter 3 provides an insight into the topic of realistic limitations when drag reducing control is applied in a DNS of TCF. In terms of spatial resolution for sensors or actuators it is shown that control remains efficient if certain dimensions are not exceeded. These dimensions are linked to the characteristic dimensions of turbulent structures in the near-wall region of the flow and hence can be roughly estimated by integral length scales of the near-wall region. The control scheme has to be able to capture the information of these scales (sufficient sensor resolution) and also tackle the scales on the side of the control input (sufficient actuator resolution). A fast deterioration of the control performance occurs when sensors or actuators cannot resolve the required scales properly due to inappropriate or erroneous control input. A sparse distribution of the actuator elements on the wall surface is shown to strongly influence the achievable drag reduction rates. Thus, the control efficiency is proportional to the ratio of the activated control area to the entire wall area.

Similar conclusions are drawn for the temporal resolution. Sufficient sensing and actuation update frequencies are necessary to estimate the correct control input. It is shown that a significant reduction of the lower sensing frequency limit can be achieved when the convection velocity of the flow is taken into account for the estimation of control input. An influence of the sensing/actuating frequency on the flow response to noise-polluted sensor information is revealed. A strong dependency of control performance on the noise intensity is found when sensing/actuation frequency is close to the frequency based on the simulation time-step.

Investigation of control schemes with wall sensing shows various successful approaches, however none of the examined control schemes outperforms the scheme of opposition control with a sensing plane in the flow field. This is expected due to deterioration of control input exactness. Suboptimal control is the only investigated control scheme that could be applied on the entire wall surface. It also yields control efficiency close to that of the opposition control. Correlation based control schemes with and without IRF-filtering could be successfully applied only in partial configuration due to pollution of sensor information at the wall. Considering the complexity of the sensor signal processing, the correlation based control with convection of the information appears to be the most simple and convenient control scheme, since only the storage of the sensor information during certain time periods is necessary for the control input estimation. Computation of the control input for a suboptimal control scheme does not require the time history of the sensor information but utilises convolution of two distributions, which makes the signal processing time-consuming. In the case of IRF-filtering a time dependent integration has to be performed for control input estimation, which would be the most expensive way of signal processing demanding the time history of the sensor signal and convolution computation. Obviously, in reality the general applicability and convenience of a certain control scheme can vary depending on the requirements of the specific application.

Due to the universality of the near-wall flow dynamics [58] the investigated limitations are not tested again in the configuration of TBL. It is assumed that the previously examined limitations regarding spatial, temporal resolution and noise pollution are completely governed by the viscous lengthscales and hence the effects discussed for TCF are also valid for TBL. Since the viscous Reynolds number continuously increases with streamwise coordinate in the developing boundary layer, the investigation of drag reducing control in a TBL is instead linked to the spatial development of the flow and its influence on the global parameters of TBL.

4 Control of Spatially Developing Turbulent Boundary Layers

This chapter presents an investigation of several control schemes within the framework of TBL. The main focus is on the control effects linked to the inherent streamwise development of TBL resulting in streamwise transients and variation of the control effect due to the increase of the Reynolds number. Hence, only the TBL-related effects of the control are discussed in the following.

A brief description of the numerical set-up and the control configurations that are utilised are provided in Section 4.1. Section 4.2 compares opposition control and suboptimal control schemes in TCF with localised control application in TBL and draws conclusions regarding control effect similarities and differences for these flow types. Next, in Section 4.3, an investigation of the control influence on the downstream behaviour of TBL is presented. Summarising, Section 4.4 presents the conclusion on the topic of skin friction drag reducing control in TBL based on the results.

The content of Sections 4.2 and 4.4 has been previously published in [135], hence the sections contain paraphrased passages and direct quotations from the publication (highlighted with quotation marks).

4.1 Numerical Procedure

The investigation is performed using direct numerical simulations (DNS) of a zero pressure gradient TBL. Figure 4.1 depicts the schematic of the simulation domain and the localised control application. The implementation is based on a pseudo-spectral solver for incompressible boundary layer flows [125]. Fourier decomposition is utilized in the horizontal and Chebyshev discretisation in the wall-normal directions. Aliasing errors are removed by dealiasing with the 3/2-rule in the streamwise and spanwise directions. The convection and viscous terms are discretised using a third-order Runge–Kutta and Crank–Nicolson methods, respectively. The flow is bounded by the wall ($y = 0$), while the spanwise and streamwise boundary conditions are periodic. At the wall, no-slip conditions are applied except for the velocity component to which the control input is imposed. A Neumann condition for the wall-normal derivative based on Falkner–Skan–Cooke solution is utilised at the free-stream boundary of the numerical domain. This is essentially the same as a free outflow boundary. Therefore, the suction velocity and the growth rate of the boundary layer are determined as a result of computation. For a sufficiently large simulation domain in the wall-normal direction ($L_y > 2.2\delta_{99}^{\max}$ in the present simulations) the upper boundary condition does not affect the turbulent solution at the lower wall. The detailed properties of the grid resolution in the area of interest and simulation domain are summarised in Table 4.1. The chosen spatial resolution is sightly coarser than the common resolution used in the recent publications on spectral DNS of TBL. The adopted resolution provides sufficiently accurate results and can be considered an optimal trade-off between high computational costs and simulation quality. We note that the statistics of the uncontrolled reference case are in good agreement with literature data [123]. Schlatter & Örlü [124] reported that TBL forgets its past by approximately $\mathrm{Re}_\theta = 2000$ independent of the tripping mechanism. However, in the present simulations the tripping is chosen in such a way that a good description of the turbulence is already achieved further upstream, approximately at $\mathrm{Re}_\theta = 600 - 700$ ($\mathrm{Re}_\tau \approx 200$). All quantities are non-dimensionalised by the free-stream velocity, U_∞, initial displacement thickness, $\delta_{d,0}$, and kinematic viscosity, ν, if not explicitly stated otherwise.

In TBL control is applied partially in the streamwise direction, while the spanwise extension of the control area covers the total domain width (Figure 4.1).

#	grid size $N_x \times N_y \times N_z$	domain size $L_x \times L_y \times L_z$	Δx^+	resolution Δy^+	Δz^+	height $\frac{L_y}{\delta_{99}^{\max}}$
1	$512 \times 129 \times 128$	$600 \times 30 \times 34$	23.8	$0.1 - 8.2$	5.9	2.25
2	$1024 \times 257 \times 128$	$1200 \times 60 \times 34$	23.8	$0.1 - 8.2$	5.9	2.88
3	$3072 \times 301 \times 256$	$3000 \times 100 \times 120$	17.8	$0.1 - 13.3$	8.9	2.32

Table 4.1: Properties of the considered simulation configurations for TBL. Viscous lengthscale is based of the average u_τ in the turbulent region of the TBL simulation.

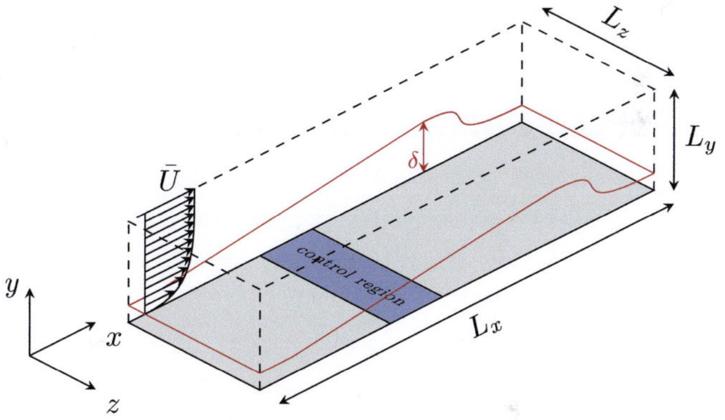

Figure 4.1: Schematic of the TBL simulation set-up.

All control areas begin at $x_0 = 186$ corresponding to $Re_\tau = 188$ as shown in Figure 2.3. Three different control areas with a streamwise extension of $\Delta x_c = 100, 150$ and 200 are introduced in TBL. The configurations are summarised in Table 4.2. Additionally, transient sections of about $\Delta x = 10 - 15$ are introduced at the beginning and at the end of the control area, in which the control amplitude is gradually increased and decreased, respectively. These transient sections are represented by a hyperbolic tangent function in order to avoid the Gibbs phenomenon [44].

	control position	control area extension
#	x_0	Δx_c
1	186	100
2	186	150
3	186	200
3	1594	200

Table 4.2: Properties of the considered control placements in TBL.

4.2 Localised Control Application

Although most turbulent flows develop in streamwise direction, direct numerical simulations (DNS) on flow control are often carried out in the simplified configuration of a fully developed turbulent channel flow (TCF) assuming that near-wall turbulent dynamics have universal features. Some more recent studies also discuss flow control schemes in developing turbulent boundary layers (TBL) and the resulting drag reduction rates are indeed similar to those found for channel flows. Drag reduction techniques that were tested in both, TCF and TBL, include opposition control through suction and blowing at the wall [18, 102], spanwise wall oscillation [80, 112, 129, 148] and uniform blowing [66, 137] where some of the TBL cases are based on a large-eddy simulations (LES) instead of a DNS.

Despite the similarity in the drag reduction rates achieved in TCF and TBL it is known that some principal differences exist in the statistical features of near-wall turbulence for these two flows even in the uncontrolled state [62]. For example, TBL exhibit stronger spanwise and wall-normal velocity fluctuations as well as stronger pressure fluctuations at the wall for the same friction Reynolds number. From the viewpoint of flow control, differences between TCF and TBL also exist when the splitting up of the skin friction drag coefficient, c_f, into its dynamical contributions is considered [37]. This so-called FIK-identity reveals that in TCF the skin friction drag can be uniquely linked to the Reynolds shear stress while additional contributions, namely one due the existence of a mean velocity in the wall-normal direction and one due to the spatial development in the streamwise direction, are present for TBL.

Since the differences described here between TCF and TBL exist, it is somewhat surprising that very similar resulting drag reduction rates are found for both flows. Control schemes for skin friction drag reduction are typically designed to reduce the Reynolds shear stress, which provides a direct link to the achieved drag reduction in TCF. For the case of uniform blowing, Kametani & Fukagata [66] and Kametani *et al.* [67] have already discussed that significant changes to the skin friction drag can also arise from TBL specific contributions to the c_f-value.

In the present section an investigation of opposition control and suboptimal control in a turbulent boundary layer is performed in order to analyse the nature of previously observed similarity in drag reduction rates [102]. The discussion is extended to the topic of the inherent differences in the statistical features present between uncontrolled TCF and TBL and their influence on the resultant control efficiency. Specifically, the control power input and the net energy saving achieved by the control schemes in TBL are reported and analysed.

4.2.1 Opposition Control

In order to perform a direct comparison between TCF and TBL at a number of different friction Reynolds numbers, five DNS of TCF (each driven by a prescribed flow rate) are carried out. The Reynolds numbers of the TCF are chosen in such a way that the friction based Reynolds numbers for the uncontrolled TCF are within the range found for the uncontrolled TBL. The TCF code utilises a finite difference method on a staggered grid with a fractional step method for pressure decoupling. The flow is bounded by the lower ($y^+ = 0$) and upper wall ($y^+ = 2\mathrm{Re}_\tau$), while periodic boundary conditions are applied in spanwise and streamwise directions. The detailed numerical scheme and its validation can be found in Stroh *et al.* [134]. A simulation domain size of $\left(L_x^+ \times L_y^+ \times L_z^+\right) = (2\pi\mathrm{Re}_\tau \times 2\mathrm{Re}_\tau \times \pi\mathrm{Re}_\tau)$ with a grid of $\left(N_x \times N_y \times N_z\right) = (160 \times 257 \times 128)$ nodes is employed. Further numerical conditions for TCF simulations are summarised in Table 4.3. An adaptive adjustment of the computational time step is utilised during the TBL simulation resulting in a mean time step of $\Delta t^+_{TBL} = 0.028$. A constant computational time step of $\Delta t^+_{TCF} = 0.014$ is used in the TCF simulations. Statistical averaging for TCF and TBL simulations is performed during 100 – 150 eddy turnover times after the controlled flow reaches an equilibrium state.

Re_τ	Δx^+	Δy^+_{min}	Δy^+_{max}	Δz^+	U_b	y^+_s
150	5.9	0.09	2.9	3.7	15.50	13.9
180	7.1	0.11	4.3	4.4	15.83	12.9
227	8.9	0.14	4.3	5.6	16.32	11.8
270	10.6	0.17	4.3	6.6	16.59	11.6
300	11.8	0.18	5.7	7.4	16.77	11.6

Table 4.3: Configuration parameters of TCF simulations. Re_τ is given for the uncontrolled flow state.

For the present TBL simulations, configuration 1 from Table 4.1 is used. This configuration enables an investigation in a friction Reynolds number range of $\mathrm{Re}_\tau = 170{-}270$. The opposition control with a wall-normal velocity component has been chosen for the localised control investigation due to the presence of reference material in TCF [15, 18, 23, 53] and TBL [102]. Opposition control is applied partially in the streamwise direction of TBL, while the spanwise extension of the control area covers the total domain width (Figure 4.1). In order to better investigate the transient control effect, control configurations 1 – 3 (Table 4.2) are introduced. In the TCF configuration control is applied to the entire area of both channel walls. Correspondingly, the control input at the wall for TCF is given by equation (2.64), while for the TBL it similarly reads

$$v(x, 0, z, t) = -\alpha \left(v(x, y_s, z, t) - \langle v(y_s, t) \rangle \right). \tag{4.1}$$

where α is a positive amplification factor. The spatial mean of the wall-normal velocity over the controlled area, that is, $\langle v(y_s, t) \rangle$ is subtracted in order to ensure a zero-net-mass-flux condition at the wall in TBL. Note that this spatial mean is exactly zero in the case of TCF due to continuity and periodicity in the streamwise and spanwise directions. Hereafter, the results obtained with $\alpha = 1.0$ are reported unless otherwise stated, although a systematic change to α is performed to investigate its effects on some control indices.

„According to previous studies [47], the resultant drag reduction rate becomes a maximum when the sensing plane is located at $y_s^+ = 15$ for $\mathrm{Re}_\tau = 180$ in TCF. In addition, it is known that this optimal location achieving the maximum drag reduction rate gradually approaches the wall with increasing Re [22]. In the case of TBL, the location of the detection plane could be changed in the streamwise direction so as to keep the same dimensionless distance from the wall in wall units, for example. Nonetheless, it is found that a detection plane parallel to the wall results in the maximum drag reduction rate among a variety of tested cases. In TBL, the local Reynolds number increases with the streamwise coordinate x, and therefore the optimal distance of the detection plane decreases in wall units, while the physical dimension of the local friction length increases with x. It is considered that these two factors cancel each other, so that the detection plane parallel to the wall becomes optimal in TBL. Accordingly, the detection plane is set to be parallel to the wall in both TCF and TBL in the present study. A series of tests with variation of the sensing plane location in the TBL revealed an optimum position at $y_s = 0.54$ ($n_y = 12$), which corresponds to a wall-normal coordinate decaying from $y_s^+ = 13$ to 11.5 in wall units. It is found that the optimal position of the sensing plane in TBL is slightly closer to the wall than in TCF. The difference is especially pronounced for the low Reynolds number region, where the optimum is found at $y_s^+ = 15$ for TCF, while it is at $y_s^+ = 12.8$ for TBL. This could be attributed to the instability inherent to opposition control. It is known that opposition control becomes unstable when α in equation (4.1) is too large or the detection plane is located further away from the wall [85, 89]. As reported by Jimenez *et al.* [62],

TBL exhibit higher v_{rms} for the same nominal Re_τ. This implies that applying opposition control to TBL tends to be more unstable. Therefore, the optimal detection plane slightly closer to the wall in TBL reduces the amplitude of the control input, and thus avoids the above mentioned instability.

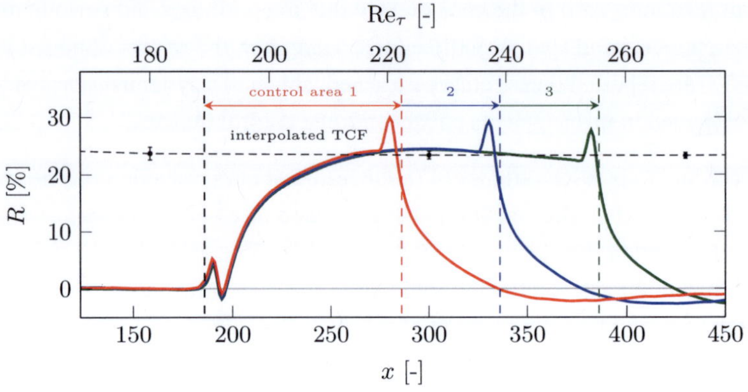

Figure 4.2: Comparison of the skin friction drag reduction distribution in TBL with interpolated controlled TCF results at $\mathrm{Re}_\tau = 150, 180, 227, 270, 300$ [135, p. 6]. Error bars represent a 3σ-confidence interval for TCF data [100].

In order to compare the results for TBL with TCF, five TCF simulations are set up in such a way that the sensing plane position as a function of the local friction Reynolds number for the uncontrolled flows and also the amplification factor ($\alpha = 1$) are the same as for TBL. As a result the sensing plane for TCF is slightly below the optimum value of $y_s^+ = 15$ (see Table 4.3)." [135, p. 4–5]

„Figure 4.2 shows the distribution of the local drag reduction rate for the three control area lengths along the streamwise coordinate within the turbulent region of the flow. A maximum local drag reduction rate of $R \approx 24\%$ is achieved. At the edges of the control area peaks in the distribution of R can be observed. These peaks are linked to strong pressure fluctuations at the wall caused by the change of boundary conditions due to control activation, in spite of the application of transient sections at the control edges." [135, p. 6]

Figure 4.3 shows the influence of the control amplitude on the distribution of drag reduction rate for the configuration with the longest control region. „It is found that the control amplitude directly affects the peak magnitude and for $\alpha < 0.4$ the peaks are not visible any more; however, the achieved drag reduction rate is also reduced.

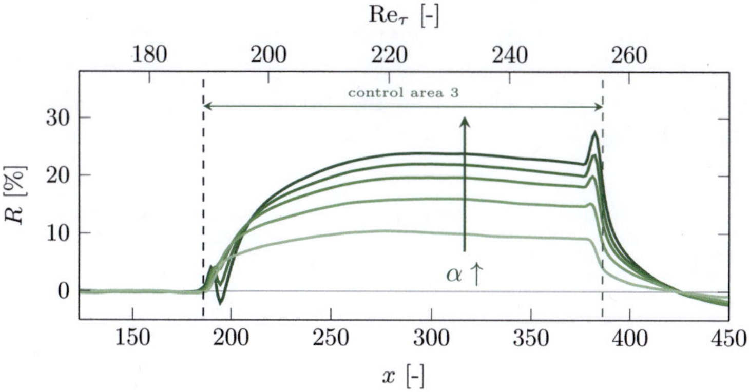

Figure 4.3: Variation of skin friction drag reduction distribution in TBL for different control amplitudes ($\alpha = 0.2, 0.4, 0.6, 0.8, 1.0$).

The drag reduction rate gradually increases in the first part of the control area and reaches a saturated state after 120 – 140 with $R \approx 24\%$. In Figure 4.2, the TCF data at different Reynolds numbers are also plotted and they are interpolated within the same range of Reynolds numbers. The presented error bars for the TCF simulations are based on the uncertainty estimation method proposed by Oliver *et al.* [100] and correspond to a 3σ-confidence interval (99.7%). It is confirmed that R achieved in TCF and TBL agree fairly well. It should also be noted that this value is comparable with previous results obtained by Chang *et al.* [15] in a TCF.

After the saturated state is reached, R exhibits a slight decrease after $x \approx 300$ for control areas 2 and 3. This behaviour is probably linked to the streamwise increase of the Reynolds number, since it is known from TCF that R decreases with Re_τ. After the control area, a relaxation section can be observed, where the flow reaches the uncontrolled state about 50 – 70 or 2000 – 3000 viscous length units downstream. This section is shorter in comparison with the results of the partial opposition control applied in turbulent pipe flow at $\mathrm{Re}_\tau = 180$, where a recovery region of 4000 – 5000 viscous length units is observed [35]. This difference might be attributed to the stronger turbulence intensity in TBL as discussed later.

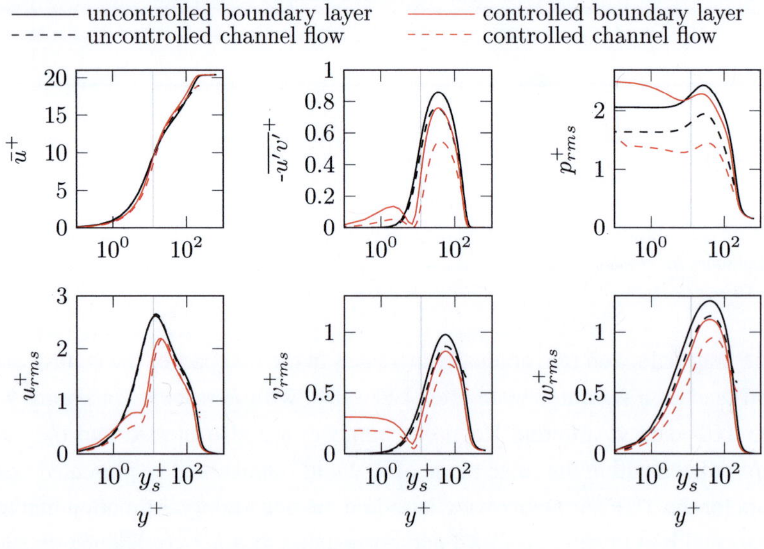

Figure 4.4: Comparison of statistical properties of uncontrolled and controlled solutions for TCF and TBL at $\mathrm{Re}_\tau = 227$ ($x = 300$) [135, p. 7].

Figure 4.4 shows a comparison of statistical properties for TBL at the streamwise position $x = 300$ with properties of TCF at the same friction Reynolds number, $\mathrm{Re}_\tau = 227$. As reported by Jimenez *et al.* [62] the uncontrolled TBL shows higher v_{rms} and w_{rms} as well as a higher Reynolds shear stress, $-\overline{u'v'}$, compared with the uncontrolled TCF. Therefore, at the position of the sensing plane ($y_s^+ = 12$) the sensed quantity, namely the instantaneous wall normal velocity component, is larger in TBL, so that the resultant control input, that is, wall blowing/suction, is also enhanced for constant α. Specifically, $v_{\mathrm{rms}}^{TCF} = 0.19$, whereas $v_{\mathrm{rms}}^{TBL} = 0.30$ (see also v_{rms}^+ at the wall in Figure 4.4). Despite the stronger control input the reduction of the Reynolds shear stress $-\overline{u'v'}^+$ is less pronounced for TBL. In addition, the controlled TBL shows enhancement of u_{rms} and p_{rms} in the near-wall region ($y^+ < 5$), which cannot be observed for TCF. Based on a parametric study with variation of the control amplitude, α, it is confirmed that this difference is simply caused by the different strengths of the control input. By reducing α below 0.4, the first peaks of u_{rms} and p_{rms} for TBL near the wall are found to disappear entirely. It is also confirmed that reducing α from unity does not further reduce the Reynolds shear stress, and the drag reduction rate becomes maximum when $\alpha = 1.0$ in TBL. Higher amplitudes ($\alpha > 1.0$) introduce oscillations into the control loops which eventually leads to the breakdown of the simulation due to increasing CFL values in the near-wall region.

The differences in the statistical properties of TBL and TCF influence the achievable local energy gains as shown in Figure 4.5. At the edges of the control area the control input power, P_{in}, tends towards zero, so that the estimated G yields high absolute values in these areas, which should not be considered in the comparison with TCF. Inside the controlled area the energy gain exhibits a gradual increase up to $G \approx 9.5$, followed by a slight decrease down to $G \approx 8.5$. Thus, the local energy gain is lower than the gain in the corresponding TCF simulations where $G \approx 10 - 13$ is achieved. In this respect it is important to note that energy gain strongly depends on the sensing plane position and on the frequency at which the control is applied [134]. Since the comparison of TBL and TCF with the same sensing plane location in terms of viscous units is utilised, the slight increase in the gain for TCF for increasing Reynolds number is attributed to the gradual change in the sensing plane position resulting in a reduced control input. In order to also allow a comparison at similar control frequencies, the interval at which the control input is updated in TCF is adjusted to the mean update interval employed in the TBL (which corresponds to the time step of the simulation, Δt^+_{TBL}). The resulting lower energy gain that is found for TBL can be attributed to the higher v_{rms} and p_{rms} near the wall that reflect in an increased power input P_{in}." [135, p. 6–8]

Conventionally, the skin friction coefficient for TCF is evaluated based on the wall friction τ_w and the bulk mean velocity U_b over the full channel height, $2h$, as introduced in equations (2.32) and (2.33), whereas the skin friction coefficient in TBL is defined based on the free-stream velocity, U_∞, introduced in equation (2.47). „The fact that similar drag reduction rates are achieved for both flows despite the lower suppression of the Reynolds shear stress in TBL can be explained by additional contributions to the skin friction coefficient in TBL. According to Fukagata, Iwamoto, and Kasagi [37], the corresponding decomposition of c_f into different dynamical contributions is given by equations (2.33) and (2.50). The Reynolds shear stress contribution is obviously present for both cases while the boundary layer contribution, c_f^δ, from TBL can be compared with the laminar contribution, c_f^L, in TCF. The latter terms are not affected by the control. For TBL two additional terms, namely c_f^C and c_f^D, are present. These terms are related to the spatial development of the TBL and to a mean convection in wall-normal direction that is not present for TCF. Both terms can be influenced by the applied control.

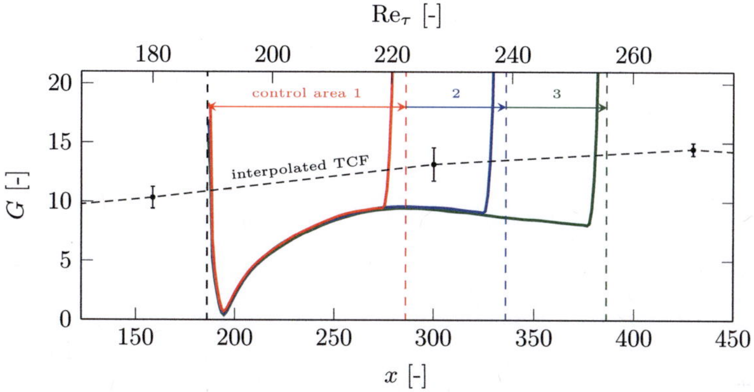

Figure 4.5: Comparison of the energy gain distribution in TBL with interpolated controlled TCF results at $\mathrm{Re}_\tau = 150, 180, 227, 270, 300$ [135, p. 7]. Error bars represent a 3σ-confidence interval for TCF data [100].

Figure 4.6 demonstrates the splitting of c_f according to equation (2.50). It can be seen that c_f^T and c_f^D are reduced in the control area, while c_f^C is increased and c_f^δ remains unchanged. Since opposition control is designed to suppress the Reynolds shear stress, the observed change in c_f^T is expected. It is rather surprising that c_f^D is reduced as much as c_f^T inside the controlled region.

Following the controlled region, the quick recovery of c_f to the uncontrolled value is mainly caused by c_f^D. The mean convection contribution, c_f^C, is negative, indicating that the wall-normal mean velocity contributes to drag reduction. In the control area its absolute value is reduced because of the smaller wall-normal flux, that is, less mass displacement.

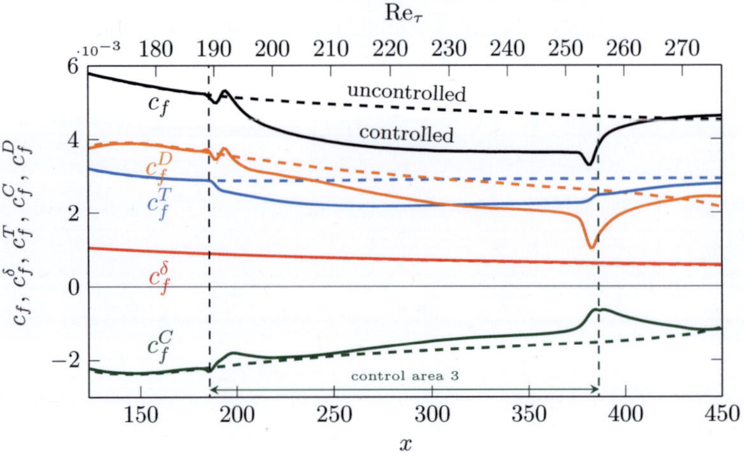

Figure 4.6: Componental contribution to the skin friction drag coefficient in TBL [135, p. 9].

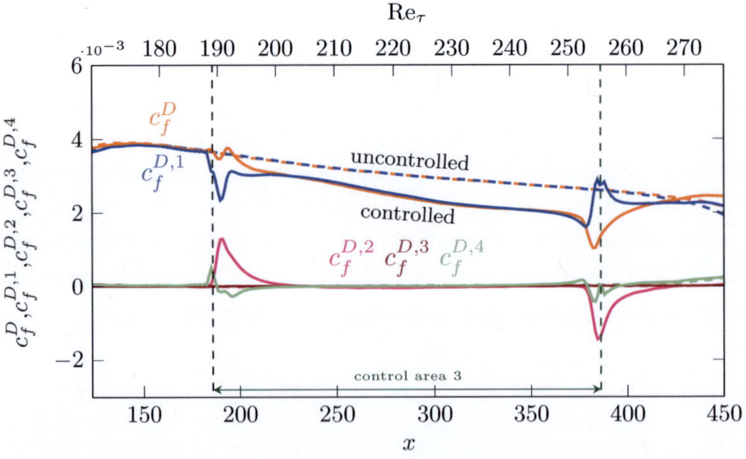

Figure 4.7: Decomposition of the spatial development contribution, c_f^D, for TBL [135, p. 9].

In order to investigate in more detail the reduction mechanisms of c_f^D in the controlled region, the four components of c_f^D are shown separately in Figure 4.7. In the uncontrolled case $c_f^{D,2}$, $c_f^{D,3}$ and $c_f^{D,4}$ are negligible and c_f^D is dominated by the streamwise gradient of the streamwise mean velocity, $c_f^{D,1}$. The constant decrease of $c_f^{D,1}$ with increasing x can generally be identified as the main cause of the well known negative gradient of c_f in the streamwise direction of the boundary layer. In the controlled case $c_f^{D,2}$ shows a relatively large value only at the beginning and the end of the control area, which oppose the variations of $c_f^{D,1}$ at these positions. The pressure term, $c_f^{D,4}$, also shows non-zero values at the edges of the control area due to the change in the boundary conditions in the streamwise direction. Inside the control area the contributions of $c_f^{D,2}$, $c_f^{D,3}$ and $c_f^{D,4}$ are negligible.

In comparison with uniform blowing in a boundary layer flow performed by Kametani & Fukagata [66], one can obviously see the different mechanism of the applied control: while the drag reduction achieved by uniform blowing occurs mainly due to the increase in the wall-normal mean convection with a significant increase of negative c_f^C, reactive control can be mainly associated with the simultaneous reduction of the Reynolds shear stress and the streamwise gradients of the flow field. The strong c_f reducing effect due to the introduction of a positive wall-normal mean velocity is also seen in the results of Pamiès *et al.* [102], where only the blowing part of the opposition control is applied.“ [135, p. 8–10]

In general, skin friction coefficient can be defined using different normalising velocity factors. Correspondingly, there exist several mathematical formulations of skin friction decomposition for the same flow configuration (see Appendix A.5). „As shown in equations (2.32) and (2.47), c_f is normalised by different velocity scales in TCF and TBL. This is linked to the fact that a quantity most relevant to an internal flow is the flow rate, whereas it is the velocity at infinite distance from a wall for an external flow. Indeed, c_f in TCF and TBL are quantitatively different even though the nominal friction Reynolds number is the same. Namely, $c_f = 7.5 \cdot 10^{-3}$ in TCF, while $c_f = 4.8 \cdot 10^{-3}$ in TBL for

$Re_\tau = 227$. In addition, the difference in normalisation is also reflected in the form and the proportional constants in the FIK identity (2.33) and (2.50), and this is an obstacle for comparing each corresponding dynamical contribution quantitatively. In order to overcome such difficulties, the centerline velocity, $U_{cl} = \bar{u}(\delta)$, can be used as a normalisation factor in TCF, which corresponds to the free-stream velocity in TBL. Accordingly, the skin friction coefficient in TCF is defined by

$$c_f = \frac{\tau_w}{0.5\rho {U_{cl}}^2}. \tag{4.2}$$

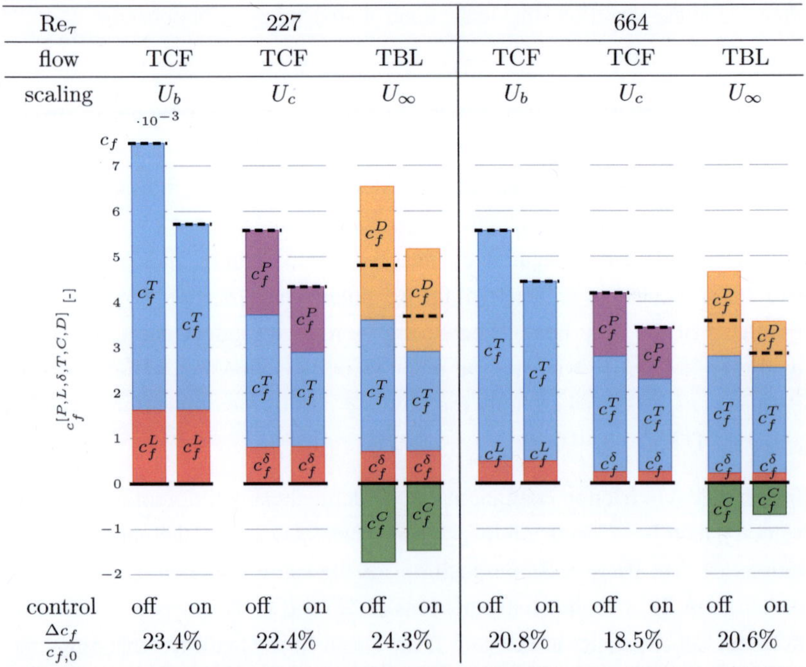

Figure 4.8: Comparison of dynamical contributions to c_f in uncontrolled and controlled TCF and TBL at $Re_\tau = 227$ and $Re_\tau = 664$ [135, p. 11].

Consequently, the following form of the FIK-identity in TCF for the newly defined c_f can be derived:

$$c_f = \underbrace{\frac{2}{3}\left(-\frac{\partial \bar{p}}{\partial x}\right)}_{\substack{c_f^P \\ \text{pressure development} \\ \text{contribution}}} + \underbrace{\frac{4(1-\delta_d)}{\mathrm{Re}_c}}_{\substack{c_f^L \\ \text{laminar} \\ \text{contribution}}} + \underbrace{4\int_0^1 (1-y)\left(-\overline{u'v'}\right)\mathrm{d}y}_{\substack{c_f^T \\ \text{Reynolds shear stress} \\ \text{contribution}}}, \tag{4.3}$$

where $\mathrm{Re}_{\mathrm{cl}} = U_{\mathrm{cl}}\delta/\nu$. The additional pressure term, c_f^P, represents the contribution from a pressure gradient that drives the flow in TCF, and originates from the spatial development contribution term $c_f^{D,4}$ in equation (2.50) for TBL. Note that, due to the momentum balance throughout the channel, c_f^P in Eq. (4.3) is exactly one third of c_f, and therefore vanishes in the original FIK identity (2.33). In the present study, c_f^P is kept, in order to compare each dynamical contribution quantitatively between TCF and TBL. The advantage of the present form (4.3) is that all the terms are similar to those in Eq. (2.50) including the multiplicative constants, and therefore the quantitative comparison of each term now becomes possible.

A comparison for opposition control in TCF and TBL is shown in Figure 4.8 where the skin friction decomposition for the uncontrolled and controlled flow states are shown at a fixed Reynolds number. For TCF, the decompositions based on U_b and U_{cl} are shown, while the contributions based on U_∞ are presented for TBL. The given Re_τ corresponds to the friction Reynolds number of the uncontrolled flows. The left part of the figure shows the results for $\mathrm{Re}_\tau = 227$." [135, p. 10]

Table 4.4 shows the values for particular FIK contributing parts based on different velocities in uncontrolled TCF and TBL. It is obvious that c_f in TBL is quantitatively close to that based on U_{cl} in TCF in comparison with the case normalised by the conventional U_b. Considering that the new FIK (4.3) for TCF has a similar form to that for TBL, and also that both c_f are normalised by the velocity scale at the outer edge of the boundary layer, the quantitative difference in c_f between TCF and TBL now contains some physical meaning. Specifically, the smaller c_f in TBL is mainly caused by a negative value of c_f^C, which is attributed to the mean wall normal velocity present only in TBL.

In the controlled flow (Table 4.5), the suppression of the turbulent contribution c_f^T is weakened in TBL and it is more pronounced in the higher Reynolds number. Meanwhile, the spatial developing term c_f^D, which is present only for TBL, is reduced by the control, so that the resultant drag reduction rate is similar in both TCF and TBL. This overall trend is essentially the same regardless of which velocity scale is used for normalisation, but the quantitative comparison between TCF and TBL becomes more straightforward by normalising the velocity scale at the outer edge of the boundary layer.

„It is observed that c_f normalised by U_{cl} instead of U_b in TCF shows better quantitative agreement with that in TBL. Its laminar and turbulent contributions in TCF and TBL are also comparable. Note that the reduction rate of c_f normalised by U_{cl} in TCF is slightly lower than that of c_f normalised by U_b in TCF or U_∞ in TBL. This is because U_{cl} is modified by the applied control, whereas U_b and U_∞ are kept constant.

Since the mathematical form of the present identity (equation (4.3)) for TCF is similar to that for TBL, the quantitative difference conveys physical meaning. For example, c_f in TBL is typically slightly smaller than that normalised by U_{cl} in TCF regardless of the presence of control, and this is primarily caused by the negative contribution of the mean convention, that is, c_f^C, which is only present in TBL. From this result, a conclusion can be made that the wall-normal momentum transfer due to the mean flow has a non-negligible impact on the wall friction in TBL. As discussed earlier, it can clearly be seen that the reduction of

c_f in TCF occurs mainly due to the attenuation of the turbulent contribution, c_f^T, while in TBL c_f is reduced due to the combined changes of three contributing components: c_f^T, c_f^C and c_f^D.

flow type	Re_τ	scale	c_f	c_f^P	$c_f^{L,\delta}$	c_f^T	c_f^C	c_f^D $\cdot 10^{-3}$
channel flow	227	U_b	7.50	-	1.61	5.89	-	-
channel flow	227	U_{cl}	5.58	1.86	0.80	2.92	-	-
boundary layer	227	U_∞	4.80	-	0.70	2.90	−1.74	2.93
channel flow	664	U_b	5.57	-	0.48	5.09	-	-
channel flow	664	U_{cl}	4.19	1.40	0.24	2.55	-	-
boundary layer	664	U_∞	3.58	-	0.21	2.58	−1.07	1.86

Table 4.4: Decomposition of the skin friction coefficient for uncontrolled TCF and TBL based on different scaling velocities. Grey background highlights similar contributing parts.

flow type	Re_τ	scale	c_f	c_f^P	$c_f^{L,\delta}$	c_f^T	c_f^C	c_f^D $\cdot 10^{-3}$
channel flow	227	U_b	5.72	-	1.61	4.11	-	-
channel flow	227	U_{cl}	4.33	1.44	0.81	2.08	-	-
boundary layer	227	U_∞	3.69	-	0.71	2.19	−1.49	2.28
channel flow	664	U_b	4.44	-	0.48	3.96	-	-
channel flow	664	U_{cl}	3.43	1.14	0.25	2.04	-	-
boundary layer	664	U_∞	2.85	-	0.21	2.32	−0.70	1.02

Table 4.5: Decomposition of the skin friction coefficient for controlled TCF and TBL based on different scaling velocities. Grey background highlights similar contributing parts.

In order to also assess the Reynolds number dependency of this result, additional simulations for TCF and TBL are carried out. The set-up of the TBL simulation utilises configuration 3 from Table 4.1 and is similar to that used by Schlatter *et al.* [125] in terms of size and resolution enabling an investigation of TBL in the region $\mathrm{Re}_\tau = 170-800$. A configuration with opposition control locally applied in the region $x = 1594-1795$ corresponding to $\mathrm{Re}_\tau = 623-674$ is considered. A corresponding TCF is simulated at $\mathrm{Re}_\tau = 664$ (for the uncontrolled case) and opposition control is then applied to both channel walls. Although the statistical integration time is limited to ten eddy turnovers, for these cases the trend in the skin friction decomposition can clearly be extracted. In the TCF at $\mathrm{Re}_\tau = 664$ the 3σ-confidence interval is estimated to be 1.4% for the drag reduction rate.

flow type	Re_τ	$\frac{\Delta c_f}{c_{f,0}}$ [%]	$\frac{\Delta c_f^P}{c_{f,0}}$ [%]	$\frac{\Delta c_f^{L,\delta}}{c_{f,0}}$ [%]	$\frac{\Delta c_f^T}{c_{f,0}}$ [%]	$\frac{\Delta c_f^C}{c_{f,0}}$ [%]	$\frac{\Delta c_f^D}{c_{f,0}}$ [%]
TCF	227	22.4	7.5	−0.2	15.2	-	-
TBL		24.3	-	−0.1	15.0	−4.9	14.1
TCF	664	18.5	6.2	−0.1	12.5	-	-
TBL		20.3	-	0	7.2	−10.4	23.5

Table 4.6: Relative changes in different dynamical contributions to c_f at $\mathrm{Re}_\tau = 227$ and $\mathrm{Re}_\tau = 664$ for TCF and TBL. Decomposition based on U_{cl} and U_∞ is utilised for TCF and TBL, respectively [135, p. 12].

The results for TBL and TCF at $\mathrm{Re}_\tau = 664$ are shown on the right-hand side of Figure 4.8. The corresponding relative change of the dynamic contributions displayed in Figure 4.8 are summarised in Table 4.6. The drag reduction achieved for TCF and TBL is surprisingly similar at both Reynolds numbers: $R \approx 22-24\%$ at $\mathrm{Re}_\tau = 227$ and $R \approx 18-20\%$ at $\mathrm{Re}_\tau = 664$. The slight deterioration in the drag reduction with increasing Reynolds number is known for TCF [41, 54]. In the case of TBL the attenuation of the turbulent contribution, c_f^T, shows a strong Reynolds number dependency in the range investigated. Its contribution to drag reduction drops from $R = 15\%$ at $\mathrm{Re}_\tau = 227$ to only $R \approx 7\%$ at $\mathrm{Re}_\tau = 664$. Interestingly, the corresponding changes in c_f^C and c_f^D lead to an increase in drag reduction and thus to a moderate total change of R when Re_τ increases from 227 to 664.

The changes in c_f^T, c_f^C and c_f^D for TBL are obviously associated with each other, which is elucidated in the following. Opposition control is designed to diminish c_f^T by suppressing quasi-streamwise vortices in the near-wall region. The reduced frictional losses at the wall lead to less mass displacement over the controlled surface such that the resulting wall-normal mean velocity is smaller than in the uncontrolled case. A smaller $\overline{v}$ results in a smaller absolute value for c_f^C, as observed in Figure 4.8. At the same time reduced frictional losses also induce weaker streamwise gradients of the streamwise mean velocity component near the wall (less momentum displacement), which results in lower values for c_f^D. The effect of control on c_f^C and c_f^D is thus governed by the reduced mass and momentum displacement when opposition control is applied along a part of the surface.

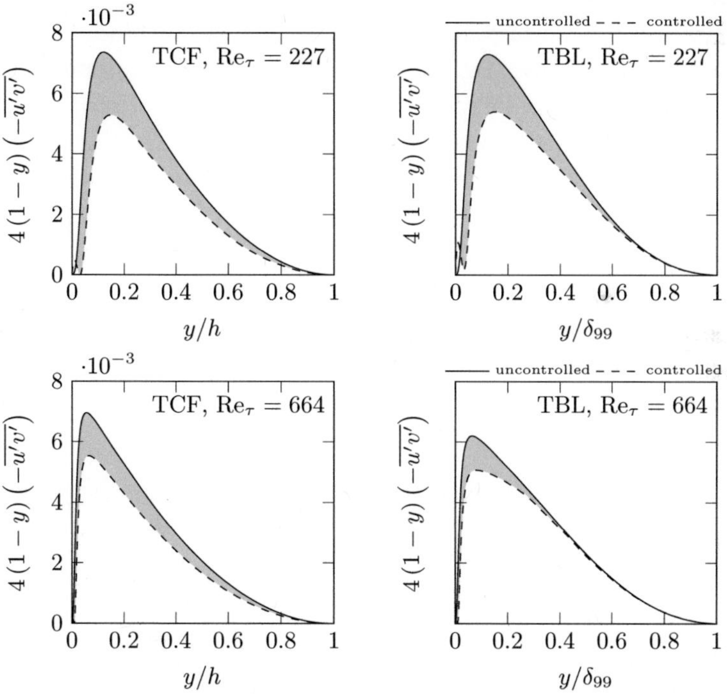

Figure 4.9: Weighted Reynolds shear stress at Re_τ = 227 and 664 for TCF and TBL. Shaded areas highlight the difference between uncontrolled and controlled states [135, p. 12].

In order to look into more details of the Reynolds number effects on the turbulent contribution, the integrands of c_f^T, that is, the weighted Reynolds shear stress, of controlled and uncontrolled flows at $\mathrm{Re}_\tau = 227$ and 664 are plotted as a function of y for TCF and TBL in Figure 4.9. The shaded areas correspond to the drag reduction contribution through c_f^T-reduction in the two flows. Obviously, the difference in reduction of the turbulent contribution between TCF and TBL is minor at $\mathrm{Re}_\tau = 227$, while the turbulent contribution away from the wall is less suppressed in TBL than in TCF at $\mathrm{Re}_\tau = 664$. Assuming that the Reynolds shear stress away from the wall is dominated by large-scale structures, the present result indicates that large-scale structures in TBL are less affected by the opposition control." [135, p. 10–12] Figure 4.10 shows the spanwise cospectra of the weighted Reynolds shear stress, $4(1 - y)(-\overline{u'v'})$, for TBL at $\mathrm{Re}_\tau = 227$ and 664. The contribution of large-scale structures (smaller wavenumbers k_z^+) to the total Reynolds shear stress is more pronounced for the higher Reynolds number away from the wall. It is also evident that application of opposition control mainly affects the near-wall region for both Reynolds numbers, while the outer region of the flow remains less affected. Thus, the result indicates a deterioration of the control effect on the large-scale motions of the flow with increasing Reynolds numbers. At the same time, however, the drag reduction effects arising from the spatial development contribution, c_f^D, and mean convection contribution, c_f^C, are more pronounced for higher Reynolds number, so that the drag reduction rate achieved in TBL still remains similar to that in TCF.

„The present results suggest that the known Re-number dependency for c_f^T in TCF [41, 54] is more pronounced for TBL, whereas the suppression of c_f^D occurs simultaneously, so that the resultant drag reduction rates in TCF and TBL are similar. As shown in Figure 4.8, the relative contribution of c_f^T to the friction drag becomes more pronounced than c_f^D with increasing Re_τ from 227 to 664. Since the suppression of c_f^D accounts for a considerable amount of the overall drag reduction for the Reynolds numbers considered here, it is of interest how the present scenario for drag reduction will be changed when the Reynolds numbers are increased further. As an indication of the behaviour at higher Reynolds numbers, the componental contributions up to $\mathrm{Re}_\tau = 2500$ (which corresponds approximately to the Reynolds number on a 4 metre car body driving at 100 km/h) are calculated from the LES database of Eitel-Amor, Örlü and Schlatter [25]. They are plotted in Figure 4.11. Comparing this figure with Figure 4.6, it is obvious that the strong variation of c_f^C and c_f^D are observed only at low Reynolds numbers, while the Reynolds number effect on the componental contributions is much weaker above $\mathrm{Re}_\tau = 664$. Based on this result, it is expected that the present scenario for drag reduction does not change significantly for higher Reynolds numbers. Meanwhile, the fact that drag reduction in TBL is achieved through the interaction of different dynamic contributions might eventually lead to different drag reduction rates for TCF and TBL.“ [135, p. 12–13]

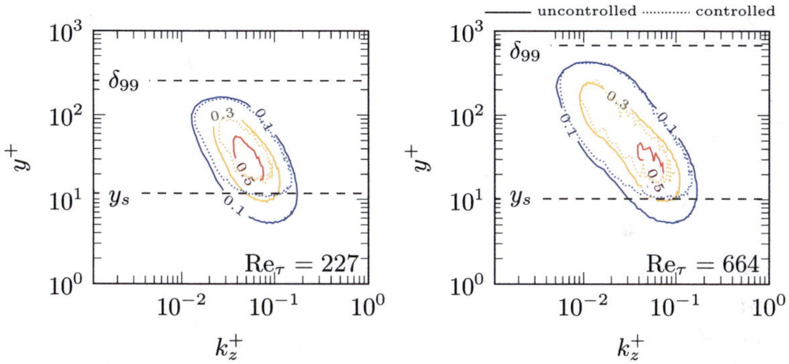

Figure 4.10: Spanwise cospectra of the weighted Reynolds shear stress at $\mathrm{Re}_\tau = 227$ and 664.

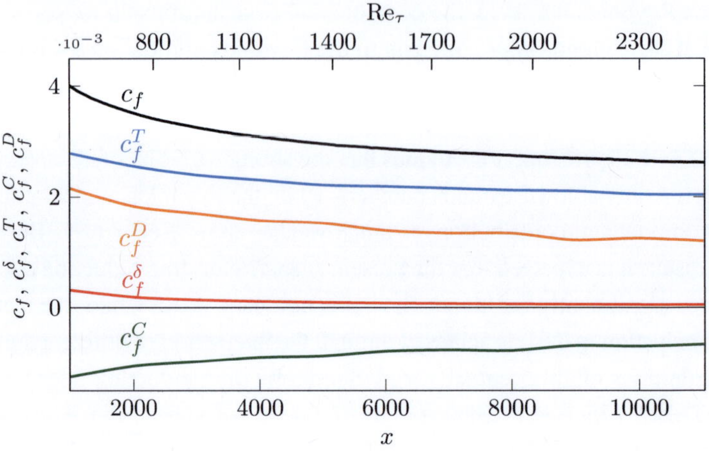

Figure 4.11: Componental contribution to the skin friction drag coefficient in uncontrolled TBL up to $\mathrm{Re}_\tau = 2500$ estimated from LES data of Eitel-Amor, Örlü and Schlatter [25], [135, p. 6].

4.2.2 Suboptimal Control

Similar to the previous section, an investigation of a suboptimal control scheme within the framework of TBL is considered in the following. The aim of the investigation is to verify the ability of a suboptimal control scheme to achieve considerable drag reduction rates based on the wall information in TBL. Simulation domain configuration 1 from Table 4.1 with control placement 3 from Table 4.2 ($x_0 = 186$, $\Delta x_c = 200$) are utilised for the simulations. Two previously discussed approaches (see Sections 2.4.2 and 3.5.3) are considered in the present work: the approach based on an analytic derivation presented by Lee *et al.* [83] and the derivation by Jeon & Choi [56], which represents a simplified response of the flow to a local wall-normal velocity impulse. Both derivations provide a weight distribution that can be used for the estimation of the control input based on the wall information. The weight distributions are transferred from the TCF configuration in order to enable a direct comparison to the results presented in Section 3.5.3. As reported in this Section, the control efficiency strongly depends on two control parameters: the control update frequency or time interval between updates, Δt_c, and the amplitude of the introduced control input, v_{rms}. Thus, as in the previous section, v_{rms} of the control input is fixed for the entire control region, so the control input amplitude depends on x and is determined as a results of the control loop activity. Consequently, a parametric study with a variation of these parameters is carried out. It is found that $v_{\text{rms}} = 1\%$ of U_∞ (corresponding to $v^+_{\text{rms}} = 0.25$) and the time interval between control input update $\Delta t_c = 0.25$ (corresponding to $\Delta t^+_c = 0.3$) provide the highest control efficiency in terms of drag reduction rate. Hence, all results presented in the following are obtained using these values.

Figure 4.12(a) shows the resulting local drag reduction rate for both suboptimal control schemes and compares it with the opposition control from the previous section. All distributions demonstrate similar streamwise development for all investigated control schemes: a transient section, followed by a saturated drag reduction state and a relaxation of the flow downstream of the control area. Both suboptimal control schemes result in an almost identical drag reduction distribution with a maximum of 17%, while opposition control yields a maximum drag reduction value of 24%. The values are also in a good agreement with the TCF simulation from Section 3.5.3, where $R = 17 - 19\%$ is observed. A slight difference in the observed R is presumably related to the 3σ-confidence interval of $\pm 1.1\%$ in the TCF simulation and a marginal difference in the Reynolds number. The performance of both suboptimal control schemes is similar to the opposition control performance with amplitude $\alpha = 0.6$, which results in the same control input amplitude, $v^+_{\text{rms}} \approx 0.25$.

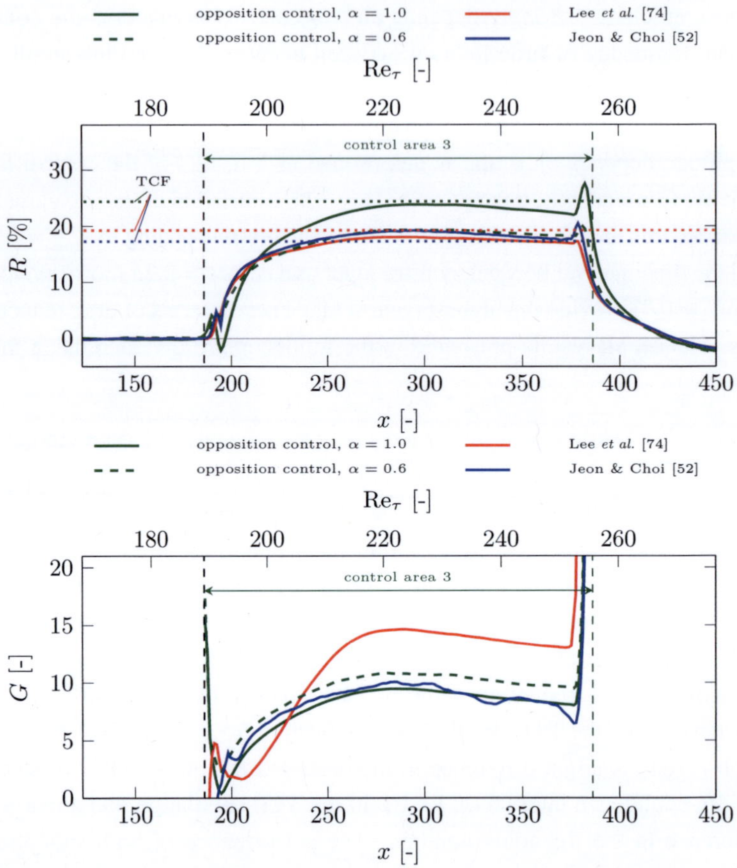

Figure 4.12: Comparison of (a) skin friction drag reduction and (b) net energy gain distribution in TBL for opposition control and two considered suboptimal control schemes. Corresponding results for TCF at $\mathrm{Re}_\tau = 180$ are plotted in (a) with dotted lines.

Figure 4.12(b) shows the development of local energy gain for the considered suboptimal control schemes and opposition control with $\alpha = 0.6$ and 1.0. The suboptimal control scheme based on the formula by Jeon & Choi [56] shows a distribution similar to one of opposition control with $\alpha = 1.0$ or 0.6, while the scheme by Lee *et al.* [83] presents a significantly different distribution. Interestingly, the energy gain of the latter scheme is smaller in the transient region close to the beginning of the control section compared with the other schemes. However, further downstream it outperforms them, providing $G = 12 - 15$, while opposition control or the scheme by Jeon & Choi [56] show $G < 10$. Comparing these results with G in TCF ($16 < G < 20$), it can be concluded that suboptimal control performs worse in TBL. As already discussed in the previous section, this difference is presumably linked to the higher p_{rms} and v_{rms} present in the TBL. Considering the integral value of G over the control area, the scheme by Lee *et al.* [83] yields the highest value among the control cases examined (Table 4.7).

	opposition control $\alpha = 1.0$	opposition control $\alpha = 0.6$	Lee *et al.* [83]	Jeon & Choi [56]
$\tilde{G}$	7.9	9.1	11.3	8.3

Table 4.7: Integral energy gain over control area.

The distribution of power input presented in Figure 4.13 reveals more details on the differences in the control loop behaviour. While, again, the P_{in} distribution of the scheme by Jeon & Choi [56] entirely resembles that of the opposition control, the scheme by Lee *et al.* [83] yields a different distribution: a strong peak in P_{in} is observed in the first third of the control area, then P_{in} gradually decreases further and falls behind the values of all the other cases. Since the power input depends mainly on p and imposed v (see equation (2.60)), the peak in P_{in} is primarily attributed to the stronger control input close to the beginning of the control area. The suboptimal scheme by Lee *et al.* [83] introduces much stronger influence in the transient section of the control region, while the control intensity further downstream is rather weak (approx. 40% of the opposition control intensity with $\alpha = 1.0$).

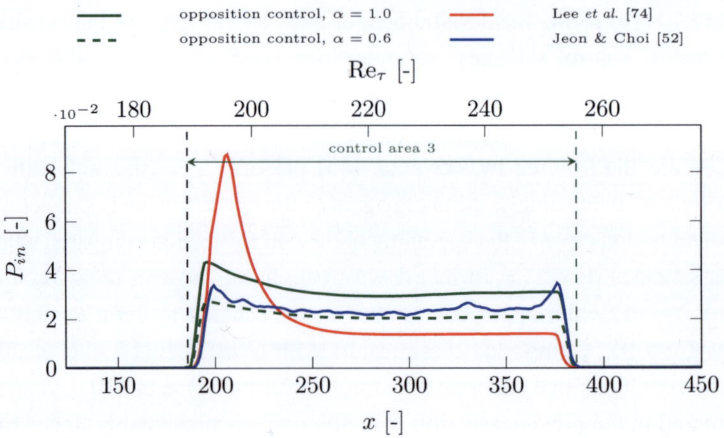

Figure 4.13: Comparison of the skin friction drag reduction distribution in TBL for opposition control and two considered suboptimal control schemes.

The investigation shows that suboptimal control schemes introduced in TCF can be directly transferred to TBL yielding very similar control efficiency in terms of drag reduction rate. Similar to opposition control, the energy gain of suboptimal control schemes applied in TBL is significantly lower than in TCF. The suboptimal control scheme based on the derivation by Lee *et al.* [83] shows better energy gain due to the redistribution of the control intensity with higher amplitudes in the beginning of control area and lower amplitudes further away. This leads to higher energy gains compared with the opposition control scheme or the suboptimal control scheme by Jeon & Choi [56].

4.3 Development Downstream of the Control Region

Although a TCF configuration has been a longstanding proven tool for the evaluation of control effects on the flow field, for many practical applications it is fairly important to manipulate external flows locally, introducing spatially transient control effects. It is difficult to reproduce these effects in TCF since the influence of any local control is eventually fed back to the inlet due to the inherent periodicity. Fukagata & Kasagi [35] investigated a partially applied opposition control in a configuration of a fully turbulent pipe flow. They reported a quick recovery of the skin friction coefficient after the control region concluding that the drag reduction achieved by partial control is proportional to the ratio of the controlled area to the total area multiplied by the drag reduction yielded for entire wall control. Recently, several attempts to transfer different control strategies proposed in a TCF to TBL have been made.

The subject of a locally applied flow control in a TBL has been rarely addressed in literature. The majority of investigations on locally applied control in TBL are conducted experimentally or analytically with a focus on applications of blowing on a partial area of wall surface. Considering numerical experiments, the following investigations have to be highlighted. Park & Choi [105] conducted a DNS of TBL with local uniform blowing or suction. This work confirms previous experiments showing a rapid decrease in the skin friction coefficient in the blowing section with subsequent enhancement of the downstream turbulent activity causing an increase in the skin friction. A similar effect was reported by Kim *et al.* [76], who investigated the effects of blowing/suction on the downstream pressure fluctuations. The effects of drag reducing uniform blowing are discussed by Kametani & Fukagata [66] and Kametani *et al.* [67] in detail based on the mathematical decomposition of the skin friction coefficient (FIK identity). Regarding localised active control, a knowledge database is even more scarce. Pamiès *et al.* [102] conducted a large eddy simulation comparing uniform blowing, opposition control and blowing-only opposition control techniques in a TBL. The paper states that the blowing-only opposition control yields about 60% drag reduction, which is significantly higher than the values achieved by the uniform blowing or the classical opposition control (less than 20%). The improvement is linked to the stronger suppression of turbulent transport towards the wall due to the positive wall-normal flux introduced by the blowing-only opposition control. These workers also discussed the transient behaviour of the manipulated TBL at the upstream and downstream edge of the control section. They observed different downstream developments resulting in a drag increase for the blowing strategies and drag reduction for the classical opposition control. However, the trends are visible only within a short region downstream of the control area; a recovery to the uncontrolled state occurs later. A recent paper by Lardeau & Leschziner [80] reported transient behaviour of TBL, where a local skin friction drag reduction is achieved by means of oscillating wall control. It was shown that the skin friction coefficient is permanently increased downstream of the control area.

In general, drag reduction rate obtained in TBL is expected to be similar to that achieved in a TCF due to the universality of near-wall turbulent dynamics. It is assumed that flow relaxation occurs quickly and the control effect rapidly decays and vanishes after a certain distance downstream of the control region. Meanwhile, it has been reported that some fundamental differences exist between uncontrolled TCF and TBL [62]. A comparison of opposition control in TCF and TBL shows that in spite of the similarity achieved for the drag reduction rate, the mechanism behind it is quite different [135]. Considering that the local state of TBL depends on the history of the upstream events, it is reasonable to assume that localised control application alters the flow state downstream. Recently, Spalart *et al.* [133] presented an analytical estimation of the drag reducing effect due to laminarisation near the leading edge of a TBL. They draw attention to the fact that one has to distinguish between local and global control effects when the spatial development of TBL is altered locally. It was shown that the alteration directly affects the global spatial development, however, a mitigation of the global effect has to be expected for a long TBL.

In the present section an investigation of localised control application in a TBL is conducted in order to examine the influence of the drag reducing control on the downstream flow development. Two characteristic control schemes with essentially different control mechanisms have been chosen for the investigation: the predetermined scheme of the uniform blowing and the scheme of body force damping. While the uniform blowing affects the convective contribution to the skin friction coefficient by introduction of additional mass flux, the body force damping scheme aims at the manipulation of the turbulent contribution representing the variety of control schemes developed for the suppression of the Reynolds shear stress. An extensive comparison of body force damping with uniform blowing is desired for the identification of differences in the drag reduction mechanisms and their influence on the downstream behaviour of the flow.

Numerical set-up 3 from Table 4.1 is chosen for the investigation, while the control placement corresponds to configuration 3 in Table 4.2 ($x_0 = 186$, $\Delta x_c = 200$). Equation (2.68) defines the control input for the uniform blowing with blowing intensity, V_w, set to 0.5% of U_∞. The reactive scheme of body force damping is based on the definition from equation (2.66) with the forcing time constant Φ fixed to $5/3$ in order to yield a drag reduction similar to the uniform blowing case. The body force is applied up to $y = 2$ such that

$$d(x,y) = \begin{cases} 1, & \text{for } 0 \leq y \leq 2 \text{ and } 186 \leq x \leq 386 \\ 0, & \text{otherwise.} \end{cases} \tag{4.4}$$

Considering viscous units based on the local wall shear stress the body force is activated in the region up to $y^+ \approx 40$. For both control schemes the control amplitude is increased and decreased smoothly within a spatial extent of $10\delta_0^*$ at the edges within the control area using a hyperbolic tangent function.

Since the aim of the present investigation is to examine the global effect of the introduced control on TBL, the following integral indices have to be considered in addition, besides the conventional definitions introduced in Section 2.4.1. In order to asses the global drag reducing effect along a plate of finite length we integrate the local skin friction coefficient in streamwise direction from the origin of the simulation domain ($x = x_s$) to a certain streamwise location x:

$$\tilde{c}_f(x) = \int_{x_s}^{x} c_f(x)\,\mathrm{d}x. \tag{4.5}$$

Correspondingly, the integral drag reduction rate provides information about the total reduction of the wall friction over a certain area of the surface:

$$\tilde{R}(x) = 1 - \frac{\tilde{c}_f(x)}{\tilde{c}_{f,0}(x)}. \tag{4.6}$$

In other words, the integral drag reduction rate represents the overall skin friction drag reduction rate achieved by a local control application in a TBL on a flat plate with length x.

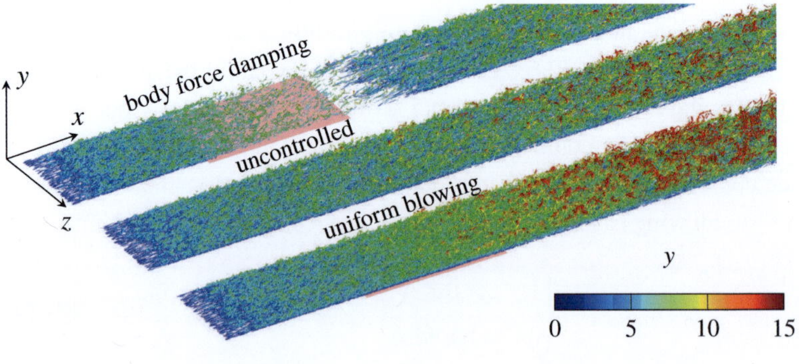

Figure 4.14: Flow structure in uncontrolled and controlled cases represented by the isosurfaces of λ_2-criterion ($\lambda_2 = -0.005$) coloured by the wall-normal coordinate. Red shaded area marks the location of the applied control.

Figure 4.14 shows the influence of the applied control on the turbulent structures of the flow. Due to cancellation of the wall-normal fluctuations in the near-wall region, a strongly pronounced attenuation of turbulent activity can be observed for body force damping. The effect is also visible over a certain area downstream of the control region, where a retransition of the flow occurs. In contrast, the application of uniform blowing rather leads to visible thickening of the TBL due to additional wall-normal mass and momentum, which is accompanied by an enhancement of turbulent activity.

4.3.1 Local and Integral Drag Reduction

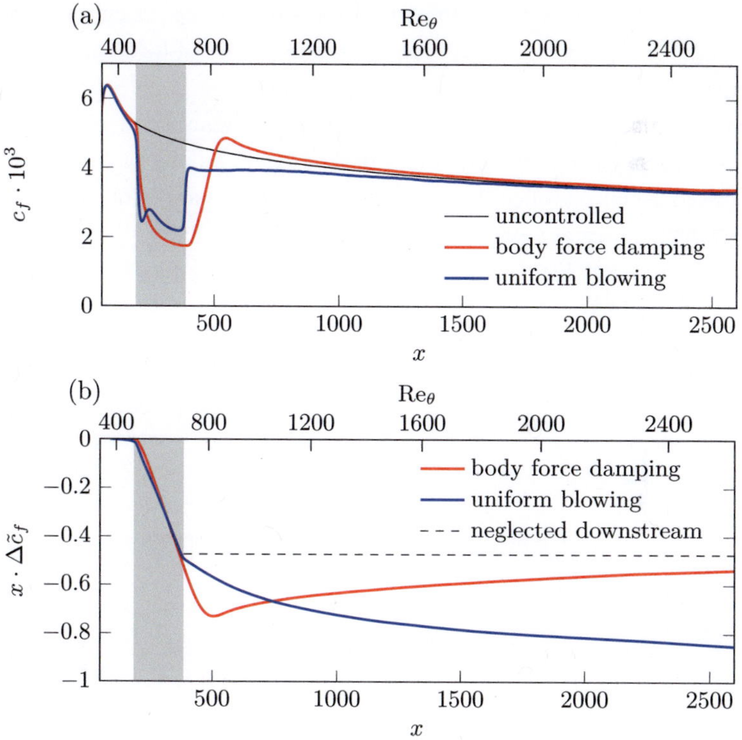

Figure 4.15: Streamwise development of (a) local skin friction coefficient and (b) difference in integral skin friction coefficient. Shaded areas mark the location of the control region.

Figure 4.15(a) depicts the development of the local skin friction coefficient for controlled and uncontrolled simulations. A substantial difference in the downstream behaviour of c_f can be observed for the manipulated flows, while a similarly strong reduction is evident in the control area. The contrast in the downstream development between body forced damping and uniform blowing is presented in Figure 4.15(b), where the difference in the integral skin friction coefficient with respect to the uncontrolled case is plotted:

$$\Delta\tilde{c}_f(x) = \tilde{c}_f - \tilde{c}_{f,0}. \tag{4.7}$$

This quantity represents the difference in the drag force acting on the plate with a length x and thus shows the cumulative effect of flow control on the global development of TBL. For comparative purposes the plot also includes the development of $\Delta\tilde{c}_f$ for the case where the differences downstream of the control are neglected entirely (black dashed line) assuming c_f immediately returns to the uncontrolled state. The uniform blowing case exhibits decreasing $\Delta\tilde{c}_f$, which indicates a permanent drag reduction, while body force damping conversely shows increasing $\Delta\tilde{c}_f$, which seems to approach the state with neglected trail effects.

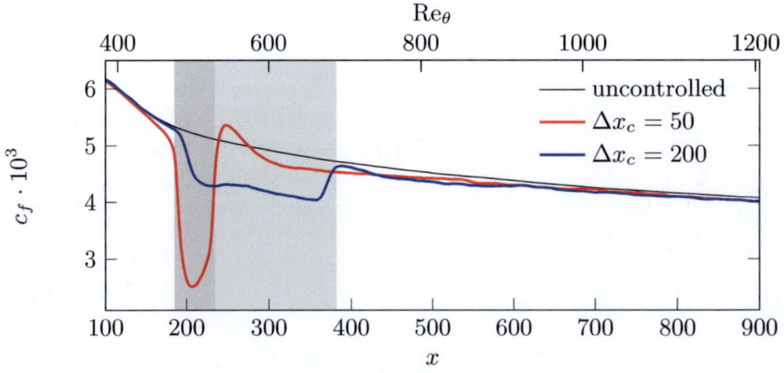

Figure 4.16: Streamwise development of local skin friction coefficient for uniform blowing with different control area lengths (Δx_c = 50, 200) and constant bulk blowing (V_w = 0.5% and 0.125%U_∞). Shaded areas mark the location of the control region.

At first sight, the downstream evolution of c_f in the case of the uniform blowing contradicts the observations that can be found in the literature. For instance,

Park & Choi [105] and Kim *et al.* [76] reported an increase in c_f in the section downstream of a blowing slit with respect to the uncontrolled TBL. However, a closer look at the simulation configuration and control parameters reveals two important points that clarify the observed differences. Firstly, the simulation domain utilised in DNS of both investigations is fairly short and therefore cannot accommodate the entire relaxation section of the flow field after the imposed blowing. Secondly, both studies adopt relatively strong blowing (up to 9.25% of U_∞) over a short control area, which results in a slightly different local development of c_f close to the control section. In order to elucidate the effect of a strong blowing over a short control region two simulations utilising shorter domains (configuration 2 from Table 4.1) are carried out. Uniform blowing is positioned at $x_0 = 186$ with control area extensions of $\Delta x_c = 50$ and 200 employing blowing intensities of 0.5% and 0.125% of U_∞, respectively. Consequently, both control configurations apply blowing with the same flow rate. Figure 4.16 shows the resulting development of c_f for these cases in comparison with the uncontrolled solution. A peak in the development of c_f can be observed at the downstream edge of the control for the shorter area blowing. However, after a distance approximately equal to the control length ($\Delta x \approx 50$) c_f crosses the uncontrolled curve and remains smaller than the uncontrolled c_f further downstream. In the case of a longer blowing section with weaker blowing, c_f always remains smaller than in the uncontrolled case. Both cases achieve similar c_f, which is smaller than the uncontrolled c_f close to the end of the simulation domain. The investigations of Park & Choi [105] and Kim *et al.* [76] are similar to the short-area control with stronger blowing, while their simulation domains are only able to capture the local increase of c_f directly downstream of the imposed blowing.

Figure 4.17(a) shows the streamwise evolution of the local drag reduction rate for the two control schemes. Both cause significant drag reduction rates with maximum values of 63% and 55%, respectively. For body force damping, drag reduction increases gradually inside the control region and approaches the maximum of $R = 63\%$ at the end of the control region. R decays quickly after the control region and assumes negative values for $\Delta x > 120$ after the control area. For uniform blowing, the flow field is already slightly influenced upstream of the control region. R rapidly rises to 53% and drops instantly at the end of the control region. In contrast to body force damping the local drag reduction rates downstream of the control region remain positive. The local decrease in R inside the control region is in agreement with literature data and is, for example, discussed in Pamiès *et al.* [102].

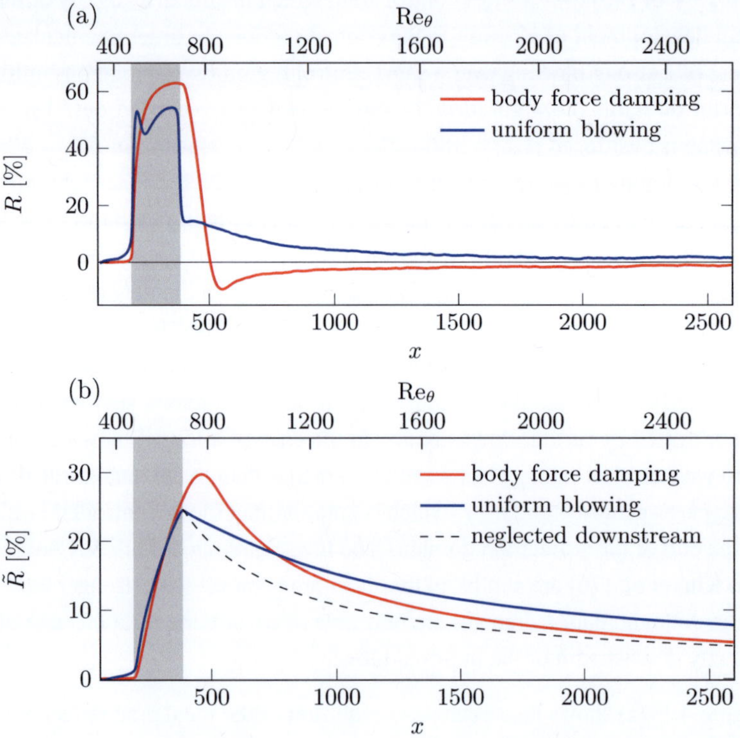

Figure 4.17: Streamwise development of (a) local drag reduction rate and (b) integral drag reduction rate. Shaded areas mark the location of the control region.

The different evolution of R downstream of the control section significantly influences the integral drag reduction rate when a longer distance of the flat plate after the control region is taken into consideration. The corresponding results for $\tilde{R}$ are shown in Figure 4.17(b). In terms of this global evaluation of drag reduction the two control schemes show very similar results until the end of the control region where both reach $\tilde{R} \approx 25\%$. If the continuation of the TBL after the control region on a flat plate of a certain length is considered, body force damping yields higher overall drag reduction rates up to a total length of $x \approx 700$. For longer plates uniform blowing provides a better integral performance.

The difference in the spatial development of R or $\tilde{R}$ is directly related to the working principle of the control techniques. The suppression of turbulence with body force damping leads to less momentum loss due to skin friction drag in the controlled region. Therefore, the boundary layer thickness increase over the control section is less than in the uncontrolled case. After the control section, the reduced boundary layer thickness remains, as a result of the control, even after retransition to a fully turbulent state has occurred. Since friction drag in TBL decreases with increasing boundary layer thickness, a locally reduced boundary layer thickness yields local higher drag. In contrast, uniform blowing directly increases the boundary layer thickness.

4.3.2 Statistics Downstream of Control Region

The spatial development of the momentum thickness for the two numerical experiments with control are shown in Figure 4.18(a) in comparison with the uncontrolled case. The previously discussed influence of the control on the boundary layer thickness can clearly be seen. The difference in momentum thickness, $\Delta\theta = \theta - \theta_0$, due to the applied control is a function of the streamwise coordinate x. It can be summarised in a single value if it is interpreted as a shift in the virtual origin of the TBL where $\Delta x_v > 0$ corresponds to a thinner TBL downstream of the control region ($\Delta\theta < 0$) and $\Delta x_v < 0$ is observed for

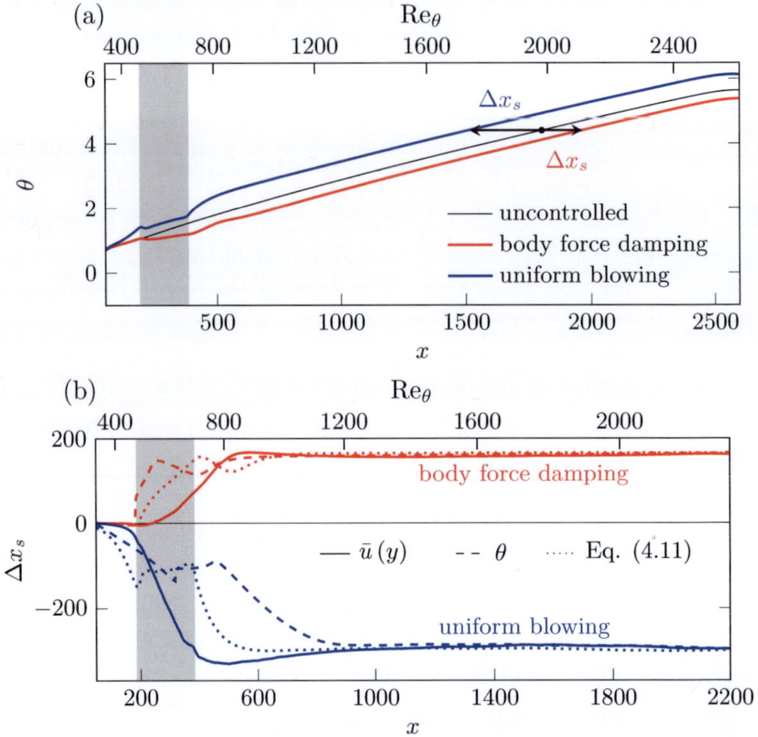

Figure 4.18: Streamwise development of (a) momentum thickness and (b) estimation of the spatial shift. Shaded areas mark the location of the control region.

a thicker TBL ($\Delta\theta > 0$). From these results, a model for TBL development subject to a local control with three regions, that is, controlled, transient and fully developed regions, can be derived as shown in Figure 4.19. When a local control is applied within a controlled region, the local drag changes depending on an applied control strategy. The controlled region is followed by a transient region, where the flow returns to an equilibrium state. Further downstream, the flow returns to the equilibrium state, so that the development of TBL is essentially the same as that of the uncontrolled flow, except for the streamwise shift Δx_v of the virtual origin of the TBL. The spatial development of the wall friction within the control and transient regions is specific to an applied control scheme, and thus has to be evaluated via either simulation or experiment. On the other hand, the wall friction in the fully developed region can be easily estimated via empirical formula without conducting expensive DNS.

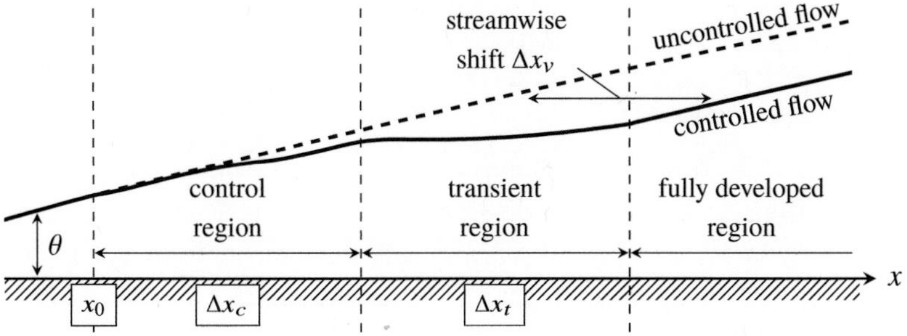

Figure 4.19: Schematic of TBL development in the case of locally applied control.

Different estimation methods for Δx_v are possible. Assuming a distribution of the mean velocity profile, $\bar{u}(x)$, or the momentum thickness is known over a long enough streamwise extent, the estimation of the spatial shift for scalar quantities $\phi(x)$ (e.g. , $\theta(x)$) can be performed searching for Δx_v where the standard deviation of the quantity becomes minimal:

$$\min\left(\phi_0(x) - \phi(x + \Delta x_v)\right)^2, \tag{4.8}$$

while the standard deviation of integral values can be used for profiles $\phi(x, y)$ (e.g. $\bar{u}(x, y)$):

$$\min\left(\int_0^\infty \phi_0(x, y)\,\mathrm{d}y - \int_0^\infty \phi(x + \Delta x_v, y)\,\mathrm{d}y\right)^2. \tag{4.9}$$

If the development in the x-direction is not known and only a set of data for a certain constant position x downstream of the control region is available, the estimation of the shift can be performed using the empirical relationship between Re_x and Re_θ proposed by Nagib *et al.* [99] based on the von Kármán integral momentum equation and Coles–Fernholz relationship [28]:

$$\mathrm{Re}_x = \frac{\mathrm{Re}_\theta}{\kappa^2}\left((\ln \mathrm{Re}_\theta + \kappa B - 1)^2 + 1\right), \tag{4.10}$$

where $\kappa = 0.387$ and $B = 4.127$. Hence, the relationship between Δx_v and θ downstream of the control region reads

$$\Delta x_v = \frac{\theta_0}{\kappa^2}\left(\ln \mathrm{Re}_{\theta,0} + \kappa B - 1\right)^2 - \frac{\theta}{\kappa^2}(\ln \mathrm{Re}_\theta + \kappa B - 1)^2 - \frac{\Delta\theta}{\kappa^2}. \tag{4.11}$$

Figure 4.18(b) shows the development of Δx_v based on the streamwise mean velocity profile, momentum thickness and empirical estimation from equation 4.11. The body force damping yields $\Delta x_v = 159$, while the uniform blowing shows $\Delta x_v = -289$ for the downstream region starting from $x = 1200 - 1400$ using the estimation based on the standard deviation. The empirical relationship provides similar values of $\Delta x_v = 165$ and 296 for body force damping and uniform blowing, respectively.

Figure 4.20 presents the statistical description of the uncontrolled and controlled cases for the same $\theta = 4.4$ corresponding to three different streamwise positions. The statistical quantities are non-dimensionalized with the local $u_\tau = \sqrt{\tau_w/\rho}$ of the particular flow. The reference position in the uncontrolled case is chosen to be at $x = 1800$, while in the controlled flows positions are chosen based on the estimated streamwise shifts yielding $x = 1959$ ($\Delta x_v = 159$) and $x = 1511$ ($\Delta x_v = -289$) for body force damping and uniform blowing, respectively. All three curves collapse, confirming the conjecture about the modification of the position of the effective virtual origin position.

4.3.3 Estimation of Downstream Behaviour

Owing to the high computational costs of TBL DNS linked to a long required streamwise extent of the numerical domain, the necessity for estimation of the flow development downstream of the control region arises. Revisiting the

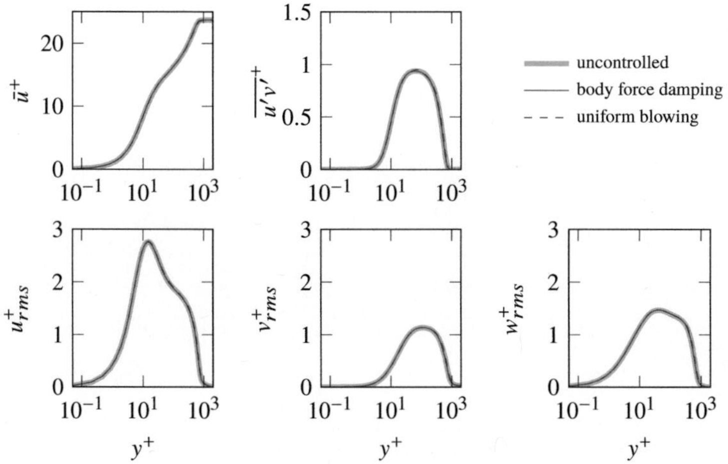

Figure 4.20: Flow statistics for the same momentum thickness (θ = 4.4) for uncontrolled and controlled cases. The momentum thickness corresponds to x = 1800, 1959 and 1511 for the uncontrolled flow, body force damping and uniform blowing, respectively.

von Kármán integral momentum equation (2.49), the relationship between the development of boundary layer momentum thickness, θ, and the skin friction coefficient is given by

$$c_f = 2\frac{\mathrm{d}\theta}{\mathrm{d}x} - \frac{2V_w}{U_\infty} - \frac{2}{\rho U_\infty^2}\int_0^\infty \frac{\partial p}{\partial x}\mathrm{d}y - \frac{2}{U_\infty^2}\int_0^\infty \frac{\partial \overline{u'u'}}{\partial x}\mathrm{d}y.$$

Integration of the equation in the streamwise direction from a certain position (e.g. , simulation domain origin $x = x_s$) leads to the following definition of the integral skin friction coefficient:

$$\tilde{c}_f(x) \approx 2\int_{x_s}^{x} \frac{\mathrm{d}\theta(x)}{\mathrm{d}x}\mathrm{d}x - 2\int_{x_s}^{x} \frac{V_w(x)}{U_\infty}\mathrm{d}x, \tag{4.12}$$

The contribution from the streamwise integral of last two terms in equation (2.49) is found to be insignificant ($< 5\%$ of $\tilde{c}_f$ for $x > 1800$ from simulation results) and is therefore neglected [123]. Assuming a block profile for $V_w(x)$ and a certain value for the momentum thickness at position x_s, the following relationship applies for the downstream flow development:

$$\tilde{c}_f(x) \approx 2\theta(x) - 2\theta(x_s) - \frac{2V_w \Delta x_c}{U_\infty} \qquad \text{for} \qquad x > x_0 + \Delta x_c. \tag{4.13}$$

Utilising this definition, the relationship between $\tilde{R}$ induced by control application and the local change in the momentum thickness can be derived from equation (4.6):

$$\tilde{R}(x) \approx -\frac{\Delta\theta(x)}{\theta_0(x) - \theta_0(x_s)} + \frac{V_w \Delta x_c}{(\theta_0(x) - \theta_0(x_s))\, U_\infty}, \tag{4.14}$$

where the second term is equal to zero for body force damping. It should also be noted that the denominator $\theta_0(x) - \theta_0(x_s)$ always remains positive as $x > x_s$. In general, the equation applies to an arbitrary drag reducing control technique.

For the body force damping a permanent positive $\Delta\theta$ is introduced, so that $\tilde{R}$ always remains positive, asymptotically approaching $\tilde{R} \to 0$ for increasing x. Uniform blowing exhibits negative $\Delta\theta$, hence the contribution from the alternation of θ always remains negative, while the contribution from the wall flux is positive and exceeds it, thus resulting in a positive integral drag reduction rate. It should be noted that both terms approach zero for increasing x.

Assuming the shift of the leading edge due to the control application is known, the downstream development of R and $\tilde{R}$ can be estimated using empirical correlation between the skin friction coefficient and the streamwise coordinate or Re_x, (e.g. , White [147]):

$$c_f = 0.4177\left(\ln\left(0.06\mathrm{Re}_x\right)\right)^{-2}. \tag{4.15}$$

A comparison of the simulation data with the estimations based on the correlation and spatial shift, Δx_v, extracted from the simulation data is presented in Figure 4.21. The proposed estimations are in a good agreement with simulation data, especially further downstream of the control, where the direct influence of the control application has almost entirely vanished and flow relaxation has occurred.

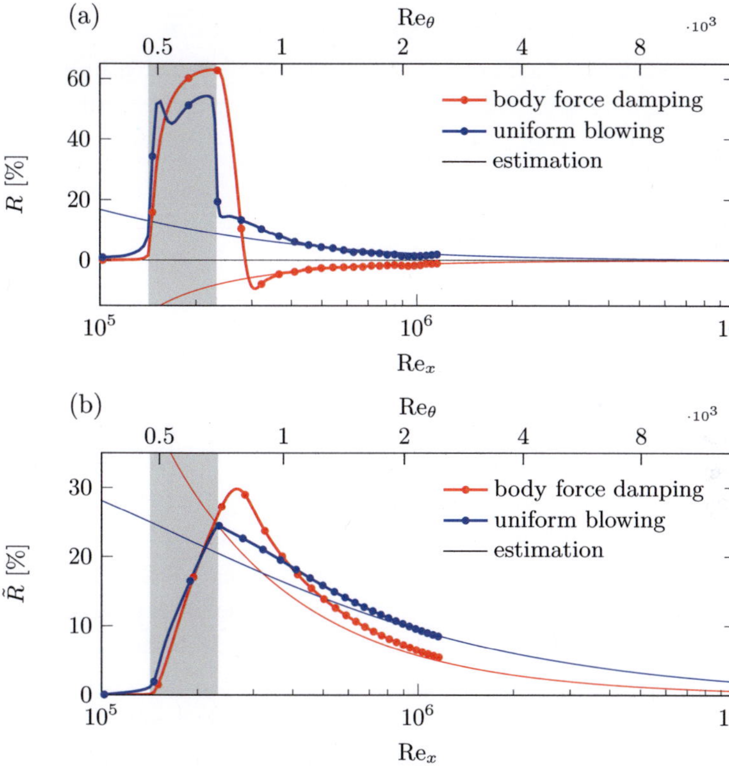

Figure 4.21: Estimation of the downstream development for (a) local drag reduction rate and (b) integral drag reduction rate based on the spatial shift. Shaded areas mark the location of the control region.

Considering the local drag reduction rate, the following relationship based on the correlation between c_f and Re_θ by Smits *et al.* [131],

$$c_f = 0.024\mathrm{Re}_\theta^{-\frac{1}{4}}, \tag{4.16}$$

can be established:

$$R(x) = 1 - \left(\frac{\theta(x)}{\theta_0(x)}\right)^{-\frac{1}{4}}. \tag{4.17}$$

Based on the relationship, an asymptotic behaviour for R can be discussed. Body force damping yields $\theta < \theta_0$ leading to positive R approaching zero for $x \to \infty$. Uniform blowing introduces $\theta > \theta_0$, so R remains strictly negative and $R \to 0$ for $x \to \infty$.

4.3.4 Influence of Control Placement

In order to analyse the influence of control placement, simulations with the streamwise position of the control region further downstream in the TBL are carried out. For these simulations the location of the control section is changed from $x_0 = 186$ (configuration 3) to $x_0 = 1594$ (configuration 4), while all other control parameters are kept the same ($\Delta x_c = 200$). Figure 4.22(a) presents a comparison of the local drag reduction rate for two considered control placements. The local control efficiency is slightly reduced for the body force damping, while it increases for uniform blowing for the control placed further downstream. Higher local drag reduction is possible in the case of uniform blowing due to the lower wall shear stress present in the downstream location, so the the effect of the imposed wall-normal velocity is stronger in comparison with the same control input placed closer to the leading edge. Lower drag reduction in the case of body force damping is attributed to the lower viscous scale present further downstream and higher local viscous Reynolds number.

Since the skin friction is not entirely governed by the small scale structures at higher Re, the control efficiency of body force damping deteriorates. For the same reasons a stronger pronounced thickening of TBL can be observed for uniform blowing, while the thinning is less pronounced for body force damping case (Figure 4.22(b)).

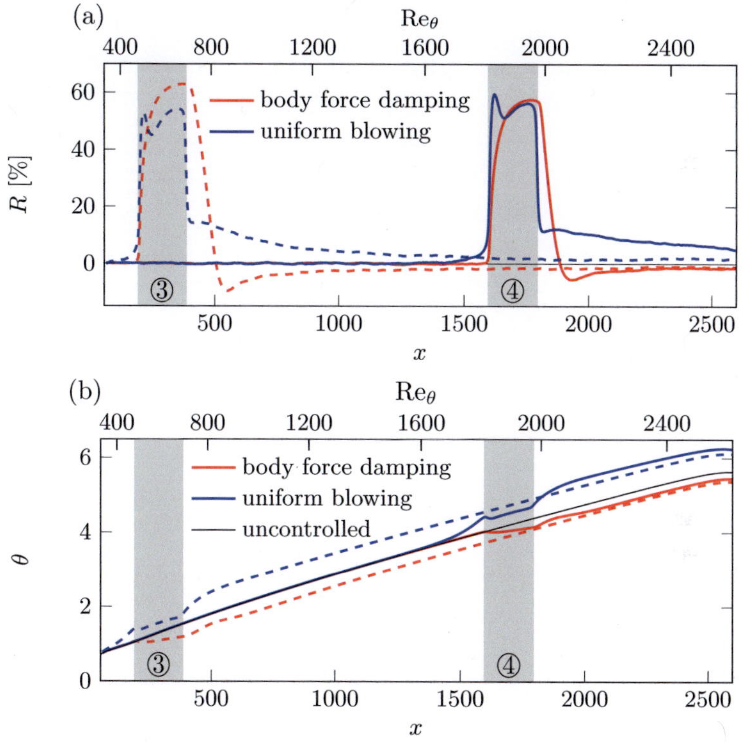

Figure 4.22: Streamwise development of (a) local drag reduction rate and (b) momentum thickness with different control placements. Shaded areas mark the location of the control regions. Configurations 3 (x_0 = 186) and 4 (x_0 = 1594) from Table 4.2 are utilised in the comparison.

Figure 4.23 presents a comparison of $\tilde{R}$ for different control placements. Body force damping exhibits lower $\tilde{R}$ for the case when control is placed further downstream due to a slightly lower local control performance linked to the increase in the local friction Reynolds number and reduction of the viscous lengthscale. Hence, the extent of the effective control region is reduced in viscous units and smaller $\Delta\theta$ are introduced downstream of the control. Uniform blowing also yields lower $\tilde{R}$ in spite of the slightly increased local control performance due to the lower local wall shear stress present downstream. Application of uniform blowing further downstream renders the TBL thicker and therefore yields a higher negative contribution from $\Delta\theta$-term from the equation (4.14). The second term is only affected by the change in θ_0, which is larger for the control placed downstream. This results in a lower contribution from the second term and leads to a lower total $\tilde{R}$. For both control schemes it is more beneficial in terms of integral drag reduction rate to apply control closer to the leading edge of the TBL.

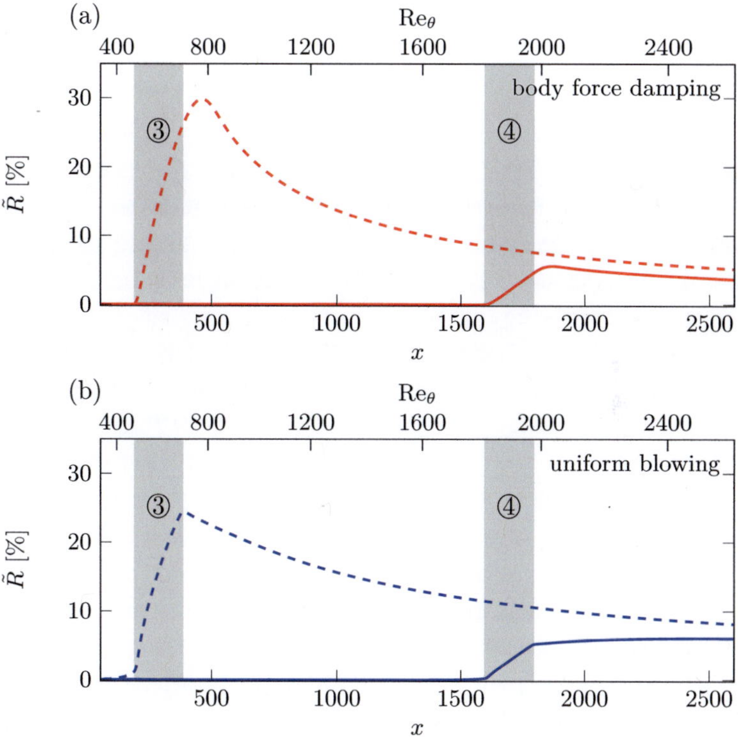

Figure 4.23: Streamwise development of the integral drag reduction rate with different control placements for (a) body force damping and (b) uniform blowing. Shaded areas mark the location of the control regions. Configurations 3 (x_0 = 186) and 4 (x_0 = 1594) from Table 4.2 are utilised in the comparison.

4.4 Concluding Remarks

„A comparison of opposition control schemes in TCF and TBL confirmed that similar drag reduction rates are achieved for both flow configurations if the control is applied at the same friction Reynolds number. Nonetheless, there are notable differences in terms of the drag reduction mechanisms and the turbulent statistics. Specifically, the higher velocity and pressure fluctuations, which are known to exist in the near-wall region of TBL, lead to higher power requirements, and therefore lower energy gains in TBL. The optimal distance of the sensing plane from the wall is found to be slightly closer in TBL. Considering the decomposition of the skin friction coefficient into its dynamical contributions, it is found that the appropriate velocity scaling (U_c for TCF and U_∞ for TBL) enables a quantitative comparison to be made between TCF and TBL regarding various contributing parts. The decomposition also reveals that the suppression of the Reynolds shear stress is weaker in TBL especially at higher Reynolds numbers, whilst there is an additional significant contribution to drag reduction that arises from changes in the streamwise gradient of the mean velocity profile. The fact that not only the attenuation of the Reynolds shear stress, but the spatial development effect inherent to TBL contributes to the overall drag reduction leads to the interesting consequence that the drag reduction rate in TBL is found to decrease only weakly with increasing Reynolds number although the suppression of the Reynolds shear stress, for which the opposition control scheme is originally designed, is significantly less pronounced.“ [135, p. 13]

An investigation of the downstream behaviour in controlled TBL using two different drag reducing control schemes shows that a flow relaxation to a different uncontrolled state occurs. The effect is entirely attributed to the mechanism of drag reduction. The reactive scheme of body force damping aimed at suppression of wall-normal velocity fluctuations attenuates turbulent activity in the control region and hence reduces skin friction drag. This leads to a slower boundary layer growth, so the TBL downstream of the control region becomes thinner in comparison with the uncontrolled flow. The effect is also expected to be similar for other drag reducing control techniques where the reduction of c_f

is linked to the attenuation of Reynolds shear stress (e.g. , oscillating walls or suboptimal control). Uniform blowing expands TBL due to the introduction of additional wall-normal momentum and reduces the wall-normal gradient of the mean velocity profile. At the same time an increase in turbulent activity in the control region is observed. A similar effect is expected to be observed for other control techniques with non-zero wall-normal mean flux such as blowing only opposition control, as proposed by Pamiès *et al.* [102]. Both control methods yield similar results in terms of local an integral drag reduction in the control region, while the downstream drag reduction is entirely different. Thinner TBL downstream of the body force damping leads to a permanent drag increase. In contrast, thicker TBL downstream of uniform blowing exhibits a persistent drag reduction. It is found that the control placement significantly influences the achievable integral drag reduction rate. Control application further from the leading edge of TBL shows a worse performance in terms of global reduction for both control schemes due to the poorer local control performance for body force damping and the stronger effect on the momentum thickness for uniform blowing.

Based on the von Kármán integral momentum equation and empirical correlation describing the development of skin friction drag, an estimation of asymptotic behaviour for an endless flat plate with localised control is proposed. The local effect of control becomes negligibly small for $x \to \infty$, as predicted by Spalart & McLean [133]. However, considering a flat plate with a limited length, quantitative estimation of the local and integral drag reduction rate can be performed assuming that the downstream behaviour of the controlled flow is described by an uncontrolled solution with a leading edge shifted upstream or downstream depending on the control method. The estimation provides good agreement with the numerical data and enables prediction of the downstream behaviour based on the knowledge of the flow state at the position of 3–4 control region lengths downstream of the control. This means that the simulation domain has to be able to accommodate only the control section and the relaxation section, while the control effect on the downstream flow field can be estimated.

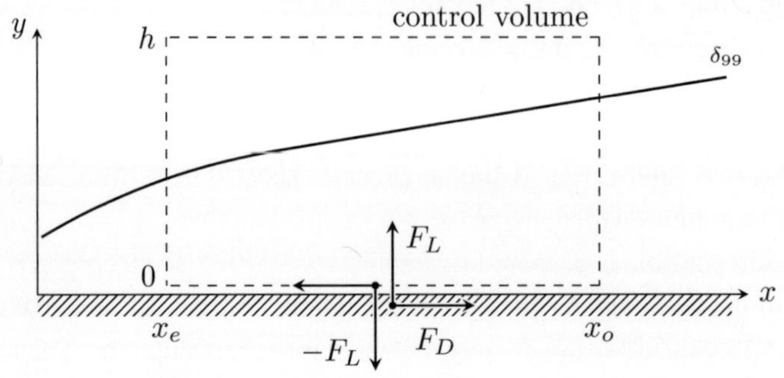

Figure 4.24: Control volume analysis of TBL flow.

It has to be emphasised that the introduced control affects not only the skin friction coefficient, which corresponds to the control effect associated with alternation of axial momentum, but also the balance of the wall-normal momentum in TBL. Considering a rectangular control volume over a flat plate with $h > \delta_{99}$ between two arbitrary streamwise locations x_e and x_o as depicted in Figure 4.24, the streamwise balance of momentum for TBL in a boundary layer approximation (see Section 2.3.2) is given by:

$$\begin{aligned} F_D &= \rho \int_0^h \bar{u}^2(x_o, y)\mathrm{d}y - \rho \int_0^h \bar{u}^2(x_e, y)\mathrm{d}y \\ &+ \int_0^h \bar{p}(x_o, y)\mathrm{d}y - \int_0^h \bar{p}(x_e, y)\mathrm{d}y, \end{aligned} \tag{4.18}$$

where F_D denotes the friction drag force per unit length due to the shear at the flat plate surface

$$F_D = \int_{x_e}^{x_o} \tau_w(x)\mathrm{d}x = \frac{1}{2}\rho U_\infty^2 \int_{x_e}^{x_o} c_f(x)\mathrm{d}x. \tag{4.19}$$

For a zero pressure gradient TBL with sufficiently large control volume the pressure terms disappear, so the drag force is completely governed by the growth of TBL reflected in the change of the streamwise velocity profile between x_e

and x_o. At the same time, the wall-normal momentum equation gives the wall-normal force imposed on the plane or lift:

$$F_L = \rho \int_{x_e}^{x_o} \bar{v}^2(x,h)\mathrm{d}x - \rho \int_{x_e}^{x_o} \bar{v}^2(x,0)\mathrm{d}x + \int_{x_e}^{x^o} \bar{p}(x,h) - \int_{x_e}^{x^o} \bar{p}(x,0), \tag{4.20}$$

where last three terms disappear for the uncontrolled TBL due to constant pressure and absence of $\bar{v}(x,0)$. This general form holds for flows where additional manipulations are introduced in the near-wall flow field. Specifically, the previously discussed schemes of opposition control or body force damping impose localised pressure gradients in the streamwise and wall-normal directions. Similarly, uniform blowing introduces streamwise pressure gradients and additional momentum due to the wall-normal velocity at the wall. Hence, the wall-normal momentum balance significantly changes, so a variation of the lift occurs depending on the applied control configuration. The control effect on the lift is of great interest for certain application scenarios, where circulation around a body or its pitching moment are of importance (e.g. airfoils). Finally, it should be noted that for a plate of finite length the overall drag is not solely determined by skin friction. Considering a control volume that surrounds a finite length plate moving through fluid, one can show that the wake of the plate, which depends on the boundary layer thickness at the trailing edge, will also contribute to the overall flow resistance. These aspects are beyond the scope of the present thesis and should be examined in future investigations.

5 Conclusion and Outlook

In the present work, an attempt is made to establish a link between drag reducing numerical control schemes developed using direct numerical simulation (DNS) of canonical flows and their potential practical application. The reference drag reducing reactive control scheme of opposition control designed for a suppression of the near-wall turbulent structures is tested with various limitations related to a realistic control application. These limitations are implemented into DNS of a fully developed turbulent channel flow (TCF) or a spatially developing turbulent boundary layer (TBL). The following limitations are considered within the thesis: finite spatial and temporal resolution of actuators/sensors, the noise polluted sensor signal as well as more realistic positioning and arrangement of the control elements. Special attention is paid to the investigation of control effects on the flow field in a configuration with a localised control application.

In contrast to the high spatial resolution of DNS, an implementation of opposition control with reduced resolution sensing and actuation is performed. The configuration is intended to resemble a more realistic control scheme with finite size actuators and a separated sensor arrangement. A parametric study with variation of actuator size, extent of separation and sensor resolution has confirmed that these parameters are strongly related to the control mechanism of opposition control. Since opposition control is developed with the intention of near-wall turbulence damping, the geometrical features of the control elements are closely linked to the spatial properties of the near-wall turbulent structures. The maximum actuator extent in the streamwise and spanwise directions as well as actuator or sensor separations correspond to the well known extents of quasi-streamwise structures ($\Delta x^+ \times \Delta z^+ \approx 200 - 400 \times 30 - 40$). Similarly, an implementation of the temporal limitations is performed followed by a parametric study, where the actuator/sensor update frequency are varied. The lower frequency limit of $f^+ = 0.2$ is identified during the survey, which corresponds to the integral time-scale of the near-wall turbulent structures ($\Delta t^+ \approx 5$). Since the influence on the control efficiency due to the introduced spatial and temporal limitations is governed by the near-wall flow dynamics, an improvement in the control performance is found when a downstream convection of sensor information is introduced. Utilising the convection velocity, $U_c^+ = 10$, a significant increase in the integral time-scale of the sensor information is observed, reducing the lower frequency limit to $f^+ = 0.04$ for the newly proposed control scheme.

Further, an introduction of artificial sensor noise in the control scheme is considered. A simultaneous variation of noise intensity and control update frequency is conducted and analysed. For the investigated range of noise intensity, drag reduction rate is found to be weakly influenced by the sensor signal pollution. However, it appears, that the net energy gain is strongly influenced by sensor noise for high frequency control due to the introduction of strong pressure fluctuations at the wall, drastically increasing control power input. Conversely, low frequency control does not exhibit a strong deterioration of energy gain with increasing intensity of the noise.

The original scheme of opposition control utilises sensor information above the wall, which constitutes an essential issue regarding practical engineering applications. Hence, an attempt is made to replace the sensors in the flow field with sensors at the wall surface. For this purpose, a series of simulations is carried out in order to analyse the spatial correlation between the original sensor signal and various flow quantities at the wall. Several quantities such as $\partial \tau_x / \partial z$ or τ_z are found to provide an appropriate replacement of the original sensor signal. However, the challenge of entire wall control evolved for the schemes with wall sensing, originating from the co-location of sensors and actuators. It is found, that in the case of co-located control elements at the wall sensor information is acutely polluted by the control activation and hence no longer provides a proper prediction of the flow state. The issue is addressed by separation of sensors from actuators resulting in an introduction of local control areas and relaxation sections, where sensors are supposed to be placed. Nonetheless, the configuration yields a significantly lower drag reduction rate ($R \approx 5\%$) than the original opposition control scheme ($R \approx 25\%$). An improvement to the control performance is enabled by activation of the sensor information filtering based on impulse response function. The filtering takes into account the flow response to the introduced control input at the wall and hence allows a minimisation of the sensor signal pollution. Application of the filtering technique yields an increase in drag reduction rate to $R \approx 10\%$. In general, it is confirmed that utilisation of wall information as the sensing quantity is very challenging.

Owing to the periodicity of the numerical domain in the configuration of TCF, an examination of local control applications is rather difficult. Any local modification of the flow field is inherently rendered on the domain inflow, so the flow configuration cannot represent a realistic flow evolution. Hence, an investigation of localised control application is carried out in a more appropriate framework of TBL, representing a broad variety of external flows known from different engineering applications.

The scheme of opposition control is transferred to the TBL simulation and examined in terms of control performance. As expected, at the same friction Reynolds number opposition control in TBL yields a drag reduction rate similar to that observed in TCF. The net energy gain is found to be slightly smaller in TBL due to the inherently stronger fluctuations of pressure and wall-normal velocity compared with TCF. However, further analysis of the control effect based on the mathematical decomposition of skin friction coefficient into its contributing parts (FIK-identity) reveals an important difference in the control mechanism. While application of opposition control in TCF leads to a suppression of the turbulent contributing part of the skin friction, application of the same control scheme in TBL also influences contributing parts linked to the spatial evolution of the layer. Interestingly, especially for higher Reynolds numbers, it is evident that the reduction of skin friction drag in TBL is more associated with the alteration of the mean convection and spatial development contribution. At the same time, the drag reduction in TCF is always attributed to the attenuation of the turbulent activity. In spite of this fact, opposition control in both flow types exhibits almost the same drag reduction rate of $R \approx 24\%$ and 20% for $\mathrm{Re}_\tau = 224$ and 664, respectively. Although a deterioration of control performance is expected for increasing Reynolds numbers, this different control mechanism found in TBL gives reason to hope that this deterioration is less pronounced than in a TCF.

A common belief that can be found in the literature conveys the statement that any type of local flow manipulation is maintained only over a short streamwise extent and a complete relaxation to the uncontrolled state eventually occurs. A previously investigated opposition control scheme provides the first insights into the alteration of TBL downstream of the control region, but the induced modification is found to be rather weak. Therefore, the topic of the control effect on the downstream development and hence the global behaviour of TBL is assessed utilising two additional control techniques that can provide higher drag reduction rates with a correspondingly stronger effect on the downstream flow state.

The reactive scheme of body force damping with a working principle similar to that of opposition control is implemented and compared with the simple and highly efficient active scheme of constant uniform blowing. Both schemes are applied locally close to the domain inlet in order to be able to observe the entire far field flow development in the downstream. Despite similar drag reduction rates in the control region ($R \approx 60\%$), the downstream behaviour of the chosen schemes varies drastically. The body force damping introduces a permanent local drag increase in the downstream section, while the uniform blowing exhibits the opposite effect. This effect is especially pronounced when an integral form of control performance indices, such as an integral skin friction coefficient or integral drag reduction rate for a flat plate with limited length, are introduced. Analysis of flow statistics has shown that this effect is linked to the control mechanism and is mainly associated with a thickening or thinning of TBL due to the control application. It is confirmed that flow relaxation to the uncontrolled state occurs in the downstream, but this uncontrolled state corresponds to a TBL with a virtual origin shifted upstream or downstream of the original uncontrolled solution, depending on the applied control scheme. Based on the induced virtual shift of the uncontrolled solution a model for estimation of the skin friction in the downstream of the control area is proposed and validated using simulation data. Estimation based on the model enables utilisation of a shorter simulation domain, since the downstream development of the skin friction coefficient becomes known.

The present thesis has treated the most obvious general limitations applicable to a broad range of numerical reactive control schemes. However, considering a particular control application or certain actuator or sensor specifications, various additional limitations will likely have to be taken into account, such as, for example, actuator shape, velocity profiles imposed by actuator activation or temporal development of the imposed velocity. Also, the definitions of net energy gain and net energy saving rate in a real control application will have to be modified with respect to the real electrical/mechanical efficiency of a chosen actuator type.

Considering future investigations, the following important topics have to be addressed:

- Clarification of the performance for near-wall turbulence control methods at technically relevant Reynolds numbers.
- Thorough analysis of the available drag reducing control schemes applied in TBL in terms of net energy gain and net energy saving rate.
- Investigation of the pressure gradients imposed by various control schemes in a TBL and quantification of its impact on the lift.
- Examination of the trailing edge effect on the total drag for a finite length plate moving through fluid.
- Investigation of possibilities for control performance improvements utilising flow specific characteristics (e.g. in TBL).

Concluding, it is hoped that the present thesis can provide a useful contribution to the available knowledge base and that potentially it will inspire a successful design or implementation of a novel control technique in numerical simulations or engineering applications.

Bibliography

[1] F. Abergel and R. Temam. On some control problems in fluid mechanics. *Theor. Comp. Fluid Dyn.*, 1(6):303–325, 1990.

[2] R. Antonia and D. K. Bisset. Spanwise structure in the near-wall region of a turbulent boundary layer. *J. Fluid Mech.*, 210:437–458, 1990.

[3] G. Batchelor. *An introduction to fluid dynamics*. Cambridge University Press, 2000.

[4] D. W. Bechert, G. Hoppe, and W. E. Reif. On the drag reduction of the shark skin. *AIAA J.*, 85:0546, 1985.

[5] D. W. Bechert, M. Bruse, W. Hage, and R. Meyer. Fluid mechanics of biological surfaces and their technological application. *Naturwissenschaften*, 87(4):157–171, 2000.

[6] T. Bewley and B. Protas. Skin friction and pressure: the "footprints" of turbulence. *Phys. D*, 196(1):28–44, 2004.

[7] T. Bewley, H. Choi, R. Temam, and P. Moin. Optimal feedback control of turbulent channel ow. *Annu. Res. Briefs, Center for Turbulence Research, Stanford University/NASA Ames*, 1993.

[8] T. Bewley, P. Moin, and R. Temam. DNS-based predictive control of turbulence: an optimal benchmark for feedback algorithms. *J. Fluid Mech.*, 447:179–225, 2001.

[9] J. Bidwell. Application of the von karman momentum theorem to turbulent boundary layers. Technical report, DTIC Document, 1951.

[10] T. J. Black, A. J. Sarnecki, and W. A. Mair. *The turbulent boundary layer with suction or injection*, volume 20. ARC, 1958.

[11] R. F. Blackwelder and H. Eckelmann. Streamwise vortices associated with the bursting phenomenon. *J. Fluid Mech.*, 94(03):577–594, 1979.

[12] E. Burger, W. Munk, and H. Wahl. Flow increase in the trans alaska pipeline through use of a polymeric drag-reducing additive. *J. Petrol. Technol.*, 34(02):377–386, 1982.

[13] J. Burgers. *Mathematical examples illustrating relations occurring in the theory of turbulent fluid motion.* Springer, 1995.

[14] A. Buri. *Eine Berechnungsgrundlage für die turbulente Grenzschicht bei beschleunigter und verzögerter Grundströmung*. PhD thesis, Diss. Techn. Wiss. ETH Zürich, Nr. 652, 1931.

[15] Y. Chang, S. Collis, and S. Ramakrishnan. Viscous effects in control of near-wall turbulence. *Phys. Fluids*, 14(11):4069–4080, 2002.

[16] H. Choi. *Turbulent Drag Reduction: Studies of Feedback Control and Flow Over Riblets*. PhD thesis, Stanford University, 1993.

[17] H. Choi, R. Temam, P. Moin, and J. Kim. Feedback control for unsteady flow and its application to the stochastic burgers equation. *J. Fluid Mech.*, 253:509–543, 1993.

[18] H. Choi, P. Moin, and J. Kim. Active turbulence control for drag reduction in wall-bounded flows. *J. Fluid Mech.*, 262:75–110, 1994.

[19] H. Choi, W.-P. Jeon, and J. Kim. Control of flow over a bluff body. *Annu. Rev. Fluid Mech.*, 40:113–139, 2008.

[20] K.-S. Choi. Near-wall structure of a turbulent boundary layer with riblets. *J. Fluid Mech.*, 208:417–458, 1989.

[21] Y. Chung and H J. Sung. Initial relaxation of spatially evolving turbulent channel flow with blowing and suction. *AIAA J.*, 39(11):2091–2099, 2001.

[22] Y. Chung and T. Talha. Effectiveness of active flow control for turbulent skin friction drag reduction. *Phys. Fluids*, Vol.23(No.2):025102, February 2011.

[23] Y. M. Chung and T. Talha. Effectiveness of active flow control for turbulent skin friction drag reduction. *Phys. Fluids*, 23(2):025102, 2011.

[24] R. J. Daniello, N. E. Waterhouse, and J. P Rothstein. Drag reduction in turbulent flows over superhydrophobic surfaces. *Phys. Fluids*, 21(8): 085103, 2009.

[25] G. Eitel-Amor, R. Örlü, and P. Schlatter. Simulation and validation of a spatially evolving turbulent boundary layer up to Re_θ=8300. *Int. J. Heat Fluid Flow*, 47:57–69, 2014.

[26] T. Endo, N. Kasagi, and Y. Suzuki. Feedback control of wall turbulence with wall deformation. *Int. J. Heat Fluid Flow*, 21(5):568–575, 2000.

[27] V. Falkner. A new law for calculating drag: The resistance of a smooth flat plate with turbulent boundary layer. *Aircr. Eng. Aerosp. Tec.*, 15(3): 65–69, 1943.

[28] H. Fernholz and P. Finleyt. The incompressible zero-pressure-gradient turbulent boundary layer: an assessment of the data. *Prog. Aerosp. Sci.*, 32(4):245–311, 1996.

[29] B. Finlayson. *The method of weighted residuals and variational principles*, volume 73. SIAM, 2013.

[30] J. Flatt. *Boundary Layer and Flow Control: Its Principles and Application*. Pergamon Press, 1961.

[31] O. Flores and J. Jiménez. Hierarchy of minimal flow units in the logarithmic layer. *Phys. Fluids*, 22(7):071704, 2010.

[32] R. F Fox, I. R. Gatland, R. Roy, and G. Vemuri. Fast, accurate algorithm for numerical simulation of exponentially correlated colored noise. *Phys. Rev. A*, 38(11):5938, 1988.

[33] B. Frohnapfel, Y. Hasegawa, and N. Kasagi. Friction drag reduction through damping of the near-wall spanwise velocity fluctuation. *Int. J. Heat Fluid Flow*, 31(3):434–441, 2010.

[34] K. Fukagata and N. Kasagi. Active control for drag reduction in turbulent pipe flow. *Eng. Turbul. Model. Exp.*, 5:607–616, 2002.

[35] K. Fukagata and N. Kasagi. Drag reduction in turbulent pipe flow with feedback control applied partially to wall. *Int. J. Heat Fluid Flow*, 24(4): 480–490, 2003.

[36] K. Fukagata and N. Kasagi. Suboptimal control for drag reduction via suppression of near-wall Reynolds shear stress. *Int. J. Heat Fluid Flow*, 25(3):341–350, 2004.

[37] K. Fukagata, K. Iwamoto, and N. Kasagi. Contribution of Reynolds stress distribution to the skin friction in wall-bounded flows. *Phys. Fluids*, 14:L73–L76, 2002.

[38] K. Fukagata, N. Kasagi, and P. Koumoutsakos. A theoretical prediction of friction drag reduction in turbulent flow by superhydrophobic surfaces. *Phys. Fluids*, 18(5):051703, 2006.

[39] K. Fukagata, M. Kobayashi, and N. Kasagi. On the friction drag reduction effect by a control of large-scale turbulent structures. *J. Fluid Sci. Technol.*, 5(3):574–584, 2010.

[40] M. Gad-el Hak. *Flow Control: Passive, Active, and Reactive Flow Management*. Cambridge University Press, 2000.

[41] D. Gatti and M. Quadrio. Performance losses of drag-reducing spanwise forcing at moderate values of the Reynolds number. *Phys. Fluids*, 25 (12):125109, 2013.

[42] F. Goldschmied. *Skin Friction of Incompressible Turbulent Boundary Layers under Adverse Pressure Gradients*. National Advisory Committee for Aeronautics, 1951.

[43] J. Gollub. Research in fluid dynamics: Meeting national needs. *Report of the US National Committee on Theoretical and Applied Mechanics*, 2006.

[44] D. Gottlieb and C.-W. Shu. On the gibbs phenomenon and its resolution. *SIAM review*, 39(4):644–668, 1997.

[45] M. S. Grewal. *Kalman filtering*. Springer, 2011.

[46] J. D. Hamilton. *Time series analysis*, volume 2. Princeton university press Princeton, 1994.

[47] E. P. Hammond, T. Bewley, and P. Moin. Observed mechanisms for turbulence attenuation and enhancement in opposition-controlled wall-bounded flows. *Phys. Fluids*, 10(9):2421–2423, 1998.

[48] C.-M. Ho and Y.-C. Tai. Micro-electro-mechanical-systems (MEMS) and fluid flows. *Annu. Rev. Fluid Mech.*, 30(1):579–612, 1998.

[49] S Hoyas and J. Jiménez. Scaling of the velocity fluctuations in turbulent channels up to Re_τ=2003. *Phys. Fluids*, 18(1):011702, 2006.

[50] S. Hoyas and J. Jiménez. Reynolds number effects on the Reynolds-stress budgets in turbulent channels. *Phys. Fluids*, 20(10):101511, 2008.

[51] L. M. Hudy, A. M. Naguib, and Humphreys W. M. Wall-pressure-array measurements beneath a separating/reattaching flow region. *Phys. Fluids*, 15(3):706–717, 2003.

[52] A. K. M. Hussain. Coherent structures - reality and myth. *Phys. Fluids*, 26(10):2816–2850, 1983.

[53] K. Iwamoto, Y. Suzuki, and N. Kasagi. Reynolds number effect on wall turbulence: toward effective feedback control. *Int. J. Heat Fluid Flow*, 23(5):678–689, 2002.

[54] K. Iwamoto, K. Fukagata, N. Kasagi, and Y. Suzuki. Friction drag reduction achievable by near-wall turbulence manipulation at high Reynolds numbers. *Phys. Fluids*, 17(1):011702–011702, 2005.

[55] A. H. Jazwinski. *Stochastic processes and filtering theory*. Courier Corporation, 2007.

[56] S. Jeon and H. Choi. Suboptimal feedback control of flow over a sphere. *Int. J. Heat Fluid Flow*, 31(2):208–216, 2010.

[57] J. Jeong, F. Hussain, W. Schoppa, and J. Kim. Coherent structures near the wall in a turbulent channel flow. *J. Fluid Mech.*, 332:185–214, 1997.

[58] J. Jiménez and P. Moin. The minimal flow unit in near-wall turbulence. *J. Fluid Mech.*, 225:213–240, 1991.

[59] J. Jiménez and A. Pinelli. The autonomous cycle of near-wall turbulence. *J. Fluid Mech.*, 389:335–359, 1999.

[60] J. Jiménez and M. Simens. Low-dimensional dynamics of a turbulent wall flow. *J. Fluid Mech.*, 435:81–91, 2001.

[61] J. Jiménez, J. C. Del Alamo, and O. Flores. The large-scale dynamics of near-wall turbulence. *J. Fluid Mech.*, 505:179–199, 2004.

[62] J. Jiménez, S. Hoyas, M. P. Simens, and Y. Mizuno. Turbulent boundary layers and channels at moderate Reynolds numbers. *J. Fluid Mech.*, 657: 335–360, 2010.

[63] S. Julier and J. Uhlmann. New extension of the kalman filter to nonlinear systems. In *AeroSense'97*, pages 182–193. International Society for Optics and Photonics, 1997.

[64] W. J. Jung, N. Mangiavacchi, and R. Akhavan. Suppression of turbulence in wall-bounded flows by high-frequency spanwise oscillations. *Phys. Fluids*, 4(8):1605–1607, 1992.

[65] R. E. Kalman. A new approach to linear filtering and prediction problems. *J. Basic Eng.-T ASME*, 82(1):35–45, 1960.

[66] Y. Kametani and K. Fukagata. Direct numerical simulation of spatially developing turbulent boundary layers with uniform blowing or suction. *J. Fluid Mech.*, 681:154–172, 2011.

[67] Y. Kametani, K. Fukagata, R. Örlü, and P. Schlatter. Effect of uniform blowing/suction in a turbulent boundary layer at moderate Reynolds number. *Int. J. Heat Fluid Flow*, Article in Press, 2015.

[68] S. Kang and H. Choi. Active wall motions for skin-friction drag reduction. *Phys. Fluids*, 12(12):3301–3304, 2000.

[69] N. Kasagi, Y. Suzuki, and K. Fukagata. Microelectromechanical systems-based feedback control of turbulence for skin friction reduction. *Annu. Rev. Fluid Mech.*, 41:231–251, 2009.

[70] S. Keshav. *A control-theoretic approach to flow control*, volume 21. ACM, 1991.

[71] J. Kim. Control of turbulent boundary layers. *Phys. Fluids*, 15(5): 1093–1105, 2003.

[72] J. Kim and T. Bewley. A linear systems approach to flow control. *Annu. Rev. Fluid Mech.*, 39:383–417, 2007.

[73] J. Kim and F. Hussain. Propagation velocity of perturbations in turbulent channel flow. *Phys. Fluids*, 5(3):695–706, 1993.

[74] J. Kim and P. Moin. Application of a fractional-step method to incompressible Navier-Stokes equations. *J. Comput. Phys.*, 59(2):308–323, 1985.

[75] J. Kim, P. Moin, and R. Moser. Turbulence statistics in fully developed channel flow at low Reynolds number. *J. Fluid Mech.*, 177:133–166, 1987.

[76] K. Kim, H. J. Sung, and M. K. Chung. Assessment of local blowing and suction in a turbulent boundary layer. *AIAA J.*, 40(1):175–177, 2002.

[77] R. B. Kinney. Skin-friction drag of a constant-property turbulent boundary layer with uniform injection. *AIAA J.*, 5(4):624–630, 1967.

[78] A. Kolmogorov. The local structure of turbulence in incompressible viscous fluid for very large Reynolds numbers. In *Dokl. Akad. Nauk SSSR*, volume 30, pages 299–303, 1941.

[79] P. Koumoutsakos. Vorticity flux control for a turbulent channel flow. *Phys. Fluids*, 11(2):248–250, 1999.

[80] S. Lardeau and M. A. Leschziner. The streamwise drag-reduction response of a boundary layer subjected to a sudden imposition of transverse oscillatory wall motion. *Phys. Fluids*, 25(7):075109, 2013.

[81] C. Lee and J. Kim. Control of the viscous sublayer for drag reduction. *Phys. Fluids*, 14(7):2523–2529, 2002.

[82] C. Lee, J. Kim, D. Babcock, and R. Goodman. Application of neural networks to turbulence control for drag reduction. *Phys. Fluids*, 9(6): 1740–1747, 1997.

[83] C. Lee, J. Kim, and H. Choi. Suboptimal control of turbulent channel flow for drag reduction. *J. Fluid Mech.*, 358:245–258, 1998.

[84] M. Lee and R. Moser. Direct numerical simulation of turbulent channel flow up to $\mathrm{Re}_\tau \approx 5200$. *J. Fluid Mech.*, 774:395–415, 2015.

[85] J. Lim and J. Kim. A singular value analysis of boundary layer control. *Phys. Fluids*, 16(6):1980–1988, 2004.

[86] L. Löfdahl and M. Gad-el Hak. MEMS applications in turbulence and flow control. *Prog. Aerosp. Sci.*, 35(2):101–203, 1999.

[87] A. Lozano-Durán and J. Jiménez. Effect of the computational domain on direct simulations of turbulent channels up to Re_τ=4200. *Phys. Fluids*, 26(1):011702, 2014.

[88] P. Luchini, M. Quadrio, and S. Zuccher. The phase-locked mean impulse response of a turbulent channel flow. *Phys. Fluids*, 18(12):121702, 2006.

[89] M. Luhar, A. S. Luhar, and B. J. McKeon. Opposition control within the resolvent analysis framework. *J. Fluid Mech.*, 749:597–626, 2014.

[90] F. Mehdi and C. M. White. Integral form of the skin friction coefficient suitable for experimental data. *Exp. Fluids*, 50(1):43–51, 2011.

[91] F. Mehdi, G. Johansson, C. White, and J. W. Naughton. On determining wall shear stress in spatially developing two-dimensional wall-bounded flows. *Exp. Fluids*, 55(1):1–9, 2014.

[92] H. S. Mickley, R. C. Ross, A. L. Squyers, and W. E. Stewart. Heat, mass, and momentum transfer for flow over a flat plate with blowing or suction. Technical report, Massachusetts Inst. of Tech., 1953.

[93] C. Min and H. Choi. Suboptimal feedback control of vortex shedding at low Reynolds numbers. *J. Fluid Mech.*, 401:123–156, 1999.

[94] Y. Mito and N. Kasagi. Dns study of turbulence modification with streamwise-uniform sinusoidal wall-oscillation. *Int. J. Heat Fluid Flow*, 19(5):470–481, 1998.

[95] P. Moin and J. Kim. Numerical investigation of turbulent channel flow. *J. Fluid Mech.*, 118:341–377, 1982.

[96] P. Moin and K. Mahesh. Direct numerical simulation: a tool in turbulence research. *Annu. Rev. Fluid Mech.*, 30(1):539–578, 1998.

[97] R. Moser and P. Moin. Direct numerical simulation of curved turbulent channel flow. Technical Report TM 85974, NASA, 1984.

[98] R. Moser, J. Kim, and N. Mansour. Direct numerical simulation of turbulent channel flow up to Re=590. *Phys. Fluids*, 11(4):943–945, 1999.

[99] H. Nagib, K. Chauhan, and P. Monkewitz. Approach to an asymptotic state for zero pressure gradient turbulent boundary layers. *Phil. Trans. R. Soc. A*, 365(1852):755–770, 2007.

[100] T. A. Oliver, N. Malaya, R. Ulerich, and R. D Moser. Estimating uncertainties in statistics computed from direct numerical simulation. *Phys. Fluids*, 26(3):035101, 2014.

[101] S. Orszag and Patterson G. Numerical simulation of three-dimensional homogeneous isotropic turbulence. *Phys. Rev. Lett.*, 28(2):76, 1972.

[102] M. Pamiès, E. Garnier, A. Merlen, and P. Sagaut. Response of a spatially developing turbulent boundary layer to active control strategies in the framework of opposition control. *Phys. Fluids*, 19(10):108102, 2007.

[103] M. Pamiès, É. Garnier, A. Merlen, and P. Sagaut. Opposition control with arrayed actuators in the near-wall region of a spatially developing turbulent boundary layer. *Int. J. Heat Fluid Flow*, 32(3):621–630, 2011.

[104] R. Panton. *Incompressible flow*. John Wiley & Sons, 2013.

[105] J. Park and H. Choi. Effects of uniform blowing or suction from a spanwise slot on a turbulent boundary layer flow. *Phys. Fluids*, 11(10): 3095–3105, 1999.

[106] J. R. Philip. Flows satisfying mixed no-slip and no-shear conditions. *Z. angew. Math. Mech.*, 23(3):353–372, 1972.

[107] U. Piomelli, P. Moin, and J. Ferziger. Large eddy simulation of the flow in a transpired channel. *J. Thermophys. Heat Transfer*, 5(1):124–128, 1991.

[108] S. P. Pope. *Turbulent flows*. Cambridge University Press, 2000.

[109] L. Prandtl. Über Flussigkeitsbewegung bei sehr kleiner Reibung. *Verhandlungen des dritten internationalen Mathematiker-Kongresses*, pages 484–491, 1904.

[110] M. Quadrio. Drag reduction in turbulent boundary layers by in-plane wall motion. *Phil. Trans. R. Soc. A*, 369(1940):1428–1442, 2011.

[111] M. Quadrio and P. Luchini. Integral space-time scales in turbulent wall flows. *Phys. Fluids*, 15(8):2219–2227, 2003.

[112] M. Quadrio and P. Ricco. Critical assessment of turbulent drag reduction through spanwise wall oscillations. *J. Fluid Mech.*, 521:251–271, 12 2004. ISSN 1469-7645.

[113] H. Rebbeck and K.-S. Choi. A wind-tunnel experiment on real-time opposition control of turbulence. *Phys. Fluids*, 18(3):035103, 2006.

[114] J. Reneaux. Overview on drag reduction technologies for civil transport aircraft. *ONERA: Tire a Part*, 153:1–18, 2004.

[115] O. Reynolds. On the dynamical theory of incompressible viscous fluids and the determination of the criterion. *Philos. Trans. Roy. Soc. London Ser. A*, pages 123–164, 1895.

[116] L. Richardson. *Weather prediction by numerical process*. Cambridge University Press, 2007.

[117] S. Robinson. Coherent motions in the turbulent boundary layer. *Annu. Rev. Fluid Mech.*, 23(1):601–639, 1991.

[118] R. Rogallo and P. Moin. Numerical simulation of turbulent flows. *Annu. Rev. Fluid Mech.*, 16(1):99–137, 1984.

[119] P. N. Romanenko and V. N. Kharchenko. The effect of transverse mass flow on heat transfer and friction drag in a turbulent flow of compressible gas along an arbitrarily shaped surface. *Int. J. Heat Fluid Flow*, 6(8): 727–738, 1963.

[120] D. Ross. Evaluation of the momentum integral equation for turbulent boundary layers. *J. Aeronaut. Sci.*, 20(7), 1953.

[121] J. P Rothstein. Slip on superhydrophobic surfaces. *Annu. Rev. Fluid Mech.*, 42:89–109, 2010.

[122] S. Satake and N. Kasagi. Turbulence control with wall-adjacent thin layer damping spanwise velocity fluctuations. *Int. J. Heat Fluid Flow*, 17(3):343–352, 1996.

[123] P. Schlatter and R. Örlü. Assessment of direct numerical simulation data of turbulent boundary layers. *J. Fluid Mech.*, 659:116–126, 2010.

[124] P. Schlatter and R. Örlü. Turbulent boundary layers at moderate Reynolds numbers: inflow length and tripping effects. *J. Fluid Mech.*, 710:5–34, 2012.

[125] P. Schlatter, R. Örlü, Q. Li, G. Brethouwer, J. Fransson, A. V. Johansson, P. H. Alfredsson, and D. S. Henningson. Turbulent boundary layers up to Re_θ=2500 studied through simulation and experiment. *Phys. Fluids*, 21(5):51702, 2009.

[126] H. Schlichting and K. Gersten. *Boundary-layer theory*. Springer, 2000.

[127] K. Schmitt, F. Durst, and P. Brunn. Drag reduction in flows by polymer additives: Clarification of some still open questions. *Arch. Appl. Mech.*, 61(2):119–132, 1991.

[128] J. Sillero, J. Jiménez, and R. Moser. One-point statistics for turbulent wall-bounded flows at reynolds numbers up to $\delta^+ \approx 2000$. *Phys. Fluids*, 25(10):105102, 2013.

[129] M. Skote. Turbulent boundary layer flow subject to streamwise oscillation of spanwise wall-velocity. *Phys. Fluids*, 23(8):081703, 2011.

[130] A. Smits and I. Marusic. Wall-bounded turbulence. *Phys. Today*, 66(9): 25, 2013.

[131] A. Smits, N. Matheson, and P. N. Joubert. Low-Reynolds-number turbulent boundary layers in zero and favorable pressure gradients. *J. Ship Res.*, 27(3):147–157, 1983.

[132] P. Spalart. Direct simulation of a turbulent boundary layer up to Re_θ=1410. *J. Fluid Mech.*, 187:61–98, 1988.

[133] P. Spalart and D. McLean. Drag reduction: enticing turbulence, and then an industry. *Philos. Trans. Roy. Soc. London Ser. A*, 369(1940): 1556–1569, 2011.

[134] A. Stroh, B. Frohnapfel, Y. Hasegawa, N. Kasagi, and C. Tropea. The influence of frequency-limited and noise-contaminated sensing on reactive turbulence control schemes. *J. Turbul.*, 13, 2012.

[135] A. Stroh, B. Frohnapfel, P. Schlatter, and Y. Hasegawa. A comparison of opposition control in turbulent boundary layer and turbulent channel flow. *Phys. Fluids*, 27(7):075101, 2015.

[136] D. Sturzebecher, S. Anders, and W. Nitsche. The surface hot wire as a means of measuring mean and fluctuating wall shear stress. *Exp. Fluids*, 31(3):294–301, 2001.

[137] Y. Sumitani and N. Kasagi. Direct numerical simulation of turbulent transport with uniform wall injection and suction. *AIAA J.*, 33(7): 1220–1228, 1995.

[138] Y. Suzuki and N. Kasagi. Evaluation of hot-wire measurements in wall shear turbulence using a direct numerical simulation database. *Exp. Therm. Fluid Sci.*, 5(1):69–77, 1992.

[139] N. Tetervin and C. Lin. *A general integral form of the boundary-layer equation for incompressible flow with an application to the calculation of the separation point of turbulent boundary layers.* National advisory committee for aeronautics, 1951.

[140] B. Toms. Some observation on the flow of linear polymer solutions through straight tubes at large Reynolds numbers. In *Proc. 1st Intl. Congr. on Rheology*, volume Vol. II, pages 135–141, 1948.

[141] C. Tropea, A. L. Yarin, and J. F. Foss. *Springer handbook of experimental fluid mechanics.* Springer Science & Business Media, 2007.

[142] N. Van Le. The von karman integral method as applied to a turbulent boundary layer. *J. Aeronaut. Sci.*, 19(9), 1952.

[143] A. Von Doenhoff and N. Tetervin. *Determination of general relations for the behavior of turbulent boundary layers.* National advisory committee for aeronautics, 1943.

[144] Th. von Kármán. On laminar and turbulent friction. *J. Appl. Math. Mech.*, 1(4):233–252, 1921.

[145] H. Von Storch and F. Zwiers. *Statistical analysis in climate research.* Cambridge University Press, 2001.

[146] M. J. Walsh. Riblets as a viscous drag reduction technique. *AIAA J.*, 21 (4):485–486, 1983.

[147] F. White. *Viscous fluid flow*. McGraw-Hill, New York, 2006.

[148] I. Yudhistira and M. Skote. Direct numerical simulation of a turbulent boundary layer over an oscillating wall. *J. Turbul.*, 9(12):1–17, 2011.

Nomenclature

Reynolds numbers

Symbol	Description
Re_b	bulk Reynolds number based on U_b and 2δ
$\mathrm{Re}_{\mathrm{cl}}$	Reynolds number based on U_{cl} and δ
$\mathrm{Re}_{\delta_{99}}$	Reynolds number based on U_∞ and δ_{99}
Re_{δ_d}	Reynolds number based on U_∞ and δ_d
Re_x	Reynolds number based on U_∞ and location x
Re_θ	Reynolds number based on U_∞ and θ
Re_τ	friction Reynolds number based on u_τ and δ (δ_{99})

Latin letters

upper case

Symbol	Description
A_{chan}	wall area of TCF domain
A_{con}	activated control area
C_t	temporal one-point correlation coefficient
C_s	spatial one-time correlation coefficient

C_{st}	spatio-temporal correlation coefficient
D	probability density function
F_D	friction drag force
F_L	lift force
G	net energy gain
H_{ij}	impulse response tensor
I	noise intensity
J	cost functional
L_s	integral lengthscale
L_t	integral timescale
L_x	streamwise domain extent
L_y	wall-normal domain extent
L_z	spanwise domain extent
N_x	amount of grid nodes in streamwise direction
N_y	amount of grid nodes in wall-normal direction
N_z	amount of grid nodes in spanwise direction
P	pumping power
P_{in}	control power input
$P_{in,v}$	control power input of wall-normal velocity imposed at the wall
$P_{in,w}$	control power input of spanwise velocity imposed at the wall
$P_{in,f}$	control power input of imposed body force
R	drag reduction rate
R'	reduced control area drag reduction rate
R_{fs}	drag reduction rate for finite actuator size
S	net energy saving rate
T	time instant

U_b	time averaged bulk mean velocity
U_c	convection velocity
U_{cl}	centerline velocity
U_∞	free-stream velocity
V_w	wall-normal velocity at the wall
W	weight function

lower case

Symbol	Description
c_f	skin friction coefficient
c_f^C	mean convection contribution to skin friction
c_f^δ	laminar boundary layer contribution to skin friction
c_f^D	spatial development contribution to skin friction
c_f^L	laminar contribution to skin friction
c_f^P	pressure development contribution to skin friction
c_f^T	turbulent contribution to skin friction
d	control distribution function
f	frequency
f_0	frequency corresponding to simulation time step Δt_0
f_a	actuation frequency
f_i	body force per unit mass
f_i^∇	non-dimensional body force per unit mass
f_s	sensing frequency
g	gravitational field strength
k	wave-number
k_x	streamwise wave-number

k_z	spanwise wave-number
n	random noise function
p	pressure
$p^\triangledown$	non-dimensional pressure
p_w	time averaged wall pressure
p_∞	free-stream/far field pressure
t	time
$t^\triangledown$	non-dimensional time
t_η	Kolmogorov time scale
u	instantaneous streamwise velocity
u_i	instantaneous velocity
$u_i^\triangledown$	non-dimensional instantaneous velocity
u^*	characteristic velocity
u_η	Kolmogorov velocity scale
u_τ	friction velocity
$\overline{u_i'u_j'}$	Reynolds stresses
v	instantaneous wall-normal velocity
w	instantaneous spanwise velocity
x	streamwise coordinate
x_0	streamwise position of control region
x_i	coordinate
$x_i^\triangledown$	non-dimensional coordinate
x_s	arbitrary position upstream of control area
x^*	characteristic length-scale
y	wall-normal coordinate
y_s	wall-normal coordinate of sensing plane position
z	spanwise coordinate

Greek letters

upper case

Symbol	Description
Δt	time interval
Δt_0	simulation time step
$\Delta t_{0,b}$	simulation time step for base simulation
Δt_a	control input update time interval
Δt_c	control update time step (when $\Delta t_a = \Delta t_s$)
Δt_s	time interval between measurements
$\Delta \theta$	difference of momentum thickness
Δx	streamwise extent
Δx_a	streamwise extent of actuator area
Δx_c	streamwise control area extent
Δx_s	streamwise extent of sensor area
Δx_t	streamwise extent of transition area
Δx_r	streamwise extent of recovery section
Δx_u	upstream shift of senor
Δx_v	streamwise shift of virtual origin of TBL
Δy_c	wall-normal control volume extent
Δz	spanwise extent
Δz_a	spanwise extent of actuator area
Δz_s	spanwise extent of sensor area
Φ	forcing time constant

lower case

Symbol	Description
α	positive amplification factor
γ	random variable/field
δ	channel half-height
δ_{99}	boundary layer thickness based on $0.99U_\infty$
δ_d	displacement thickness
$\delta_{d,0}$	initial displacement thickness
δ_ν	viscous lengthscale
ε	dissipation
ϵ	impulse amplitude
ζ	Fréchet derivative of velocity field
η	Kolmogorov lengthscale
θ	momentum thickness
μ	dynamic viscosity
ν	kinematic viscosity
ρ	density
τ	time averaged total shear stress
τ_w	time averaged wall shear stress
τ_x	streamwise wall shear stress
$\hat{\tau}_x$	filtered streamwise wall shear stress
$\tilde{\tau}_x$	mean response of streamwise wall shear stress
τ_z	spanwise wall shear stress
ϕ	random field/variable
φ	sensing quantity
$\hat{\varphi}$	sensing quantity in Fourier representation

$\bar{\phi}$	time averaged value
ϕ'	fluctuation
ϕ_{rms}	root mean square value

Mathematical symbols

Symbol	Description
δ_{ij}	Kronecker delta
δ_{ϵ}	Dirac delta function
var(. . .)	variance
cov(. . .)	covariance

Abbreviations

Symbol	Description
DNS	direct numerical simulation
CFR	constant flow rate
CPG	constant pressure gradient
FDM	finite difference method
IRF	impulse response function
TBL	spatially developing turbulent boundary layer
TCF	fully developed turbulent channel flow
QSV	quasi streamwise vortex
RMS	root mean square

List of Figures

List of Tables

A Appendix

A.1 Integral Momentum Equation

The derivation of the integral momentum equation based on the steady state two-dimensional Reynolds averaged Navier–Stokes equations with three-dimensional turbulence [42] is shown in the following. The initial equations for x- and y-component read:

$$\bar{u}\frac{\partial \bar{u}}{\partial x} + \bar{v}\frac{\partial \bar{u}}{\partial y} + \frac{1}{\rho}\frac{\partial \bar{p}}{\partial x} = \nu\left(\frac{\partial^2 \bar{u}}{\partial x^2} + \frac{\partial^2 \bar{u}}{\partial y^2}\right) - \frac{\partial \overline{u'u'}}{\partial x} - \frac{\partial \overline{u'v'}}{\partial y} - \frac{\partial \overline{u'w'}}{\partial z}, \tag{A.1}$$

$$\bar{u}\frac{\partial \bar{v}}{\partial x} + \bar{v}\frac{\partial \bar{v}}{\partial y} + \frac{1}{\rho}\frac{\partial \bar{p}}{\partial y} = \nu\left(\frac{\partial^2 \bar{v}}{\partial x^2} + \frac{\partial^2 \bar{v}}{\partial y^2}\right) - \frac{\partial \overline{u'v'}}{\partial x} - \frac{\partial \overline{v'v'}}{\partial y} - \frac{\partial \overline{v'w'}}{\partial z}, \tag{A.2}$$

$$0 = \frac{\partial \overline{u'w'}}{\partial x} - \frac{\partial \overline{v'w'}}{\partial y} - \frac{\partial \overline{w'w'}}{\partial z}, \tag{A.3}$$

with continuity equations

$$\frac{\partial \bar{u}}{\partial x} + \frac{\partial \bar{v}}{\partial y} = 0, \tag{A.4}$$

and

$$\frac{\partial u'}{\partial x} + \frac{\partial v'}{\partial y} + \frac{\partial w'}{\partial z} = 0. \tag{A.5}$$

Using following definitions

$$\tau_{xx} = \mu\frac{\partial \bar{u}}{\partial x} - \rho\overline{u'u'}, \tag{A.6}$$

$$\tau_{xy} = \mu\frac{\partial \bar{u}}{\partial y} - \rho\overline{u'v'}, \tag{A.7}$$

$$\tau_{yx} = \mu \frac{\partial \bar{v}}{\partial x} - \rho \overline{u'v'}, \tag{A.8}$$

$$\tau_{yy} = \mu \frac{\partial \bar{v}}{\partial y} - \rho \overline{v'v'}, \tag{A.9}$$

the relationship reads

$$\begin{aligned}
c_f \;=\;& 2\frac{\mathrm{d}}{\mathrm{d}x}\int_0^{\delta_t}\left(\frac{\bar{u}}{U_\infty} - \frac{\bar{u}^2}{U_\infty^2} + \frac{\tau_{xx}}{\rho U_\infty^2} - \frac{\tau_{yy}}{\rho U_\infty^2}\right)\mathrm{d}y \\
+\;& \frac{4}{U_\infty}\frac{\mathrm{d}U_\infty}{\mathrm{d}x}\int_0^{\delta_t}\left(\frac{\bar{u}}{U_\infty} - \frac{\bar{u}^2}{U_\infty^2} + \frac{\tau_{xx}}{\rho U_\infty^2} - \frac{\tau_{yy}}{\rho U_\infty^2}\right)\mathrm{d}y \\
+\;& \frac{2}{U_\infty}\frac{\mathrm{d}U_\infty}{\mathrm{d}x}\int_0^{\delta_t}\left(1 - \frac{\bar{u}}{U_\infty}\right)\mathrm{d}y \\
-\;& 2\int_0^{\delta_t}\int_{\delta_t}^{y}\left[2\left(\frac{1}{U_\infty^2}\left(\frac{\mathrm{d}U_\infty}{\mathrm{d}x}\right)^2 + \frac{1}{U_\infty}\frac{\mathrm{d}^2U_\infty}{\mathrm{d}x^2}\right)\frac{\tau_{yx}}{\rho U_\infty^2}\right. \\
&\left. + \frac{4}{U_\infty}\frac{\mathrm{d}U_\infty}{\mathrm{d}x}\frac{\partial}{\partial x}\left(\frac{\tau_{yx}}{\rho U_\infty^2}\right) + \frac{\partial^2}{\partial x^2}\left(\frac{\tau_{yx}}{\rho U_\infty^2}\right)\right]\mathrm{d}y\mathrm{d}y \\
-\;& \frac{4}{U_\infty}\frac{\mathrm{d}U_\infty}{\mathrm{d}x}\int_0^{\delta_t}\int_{\delta_t}^{y}\left[\frac{\partial}{\partial x}\left(\frac{\bar{u}\bar{v}}{U_\infty^2}\right) - 2\frac{\bar{v}}{U_\infty}\frac{\partial}{\partial x}\left(\frac{\bar{u}}{U_\infty}\right)\right]\mathrm{d}y\mathrm{d}y \\
-\;& 2\int_0^{\delta_t}\int_{\delta_t}^{y}\left[\frac{\bar{u}}{U_\infty}\frac{\partial^2}{\partial x^2}\left(\frac{\bar{v}}{U_\infty}\right) - \frac{\bar{v}}{U_\infty}\frac{\partial^2}{\partial x^2}\left(\frac{\bar{u}}{U_\infty}\right)\right]\mathrm{d}y\mathrm{d}y,
\end{aligned} \tag{A.10}$$

or in the short form:

$$c_f = 2\frac{\mathrm{d}\theta_t}{\mathrm{d}x} + \frac{2}{U_\infty}\frac{\mathrm{d}U_\infty}{\mathrm{d}x}(2\theta_t + \delta_d) - A - B, \tag{A.11}$$

where

$$\theta_t = \int_0^{\delta_t}\frac{\bar{u}}{U_\infty}\left[1 - \frac{\bar{u}}{U_\infty}\left(1 + \frac{\tau_{yy}}{\rho\bar{u}^2} - \frac{\tau_{xx}}{\rho\bar{u}^2}\right)\right]\mathrm{d}y, \tag{A.12}$$

$$\delta_d = \int_0^{\delta_t}\left[1 - \frac{\bar{u}}{U_\infty}\right]\mathrm{d}y, \tag{A.13}$$

and correction factors

$$\begin{aligned}
A \;=\;& 2\int_0^{\delta_t}\int_{\delta_t}^{y}\left[2\left(\frac{1}{U_\infty^2}\left(\frac{\mathrm{d}U_\infty}{\mathrm{d}x}\right)^2 + \frac{1}{U_\infty}\frac{\mathrm{d}^2U_\infty}{\mathrm{d}x^2}\right)\frac{\tau_{yx}}{\rho U_\infty^2}\right. \\
&\left. + \frac{4}{U_\infty}\frac{\mathrm{d}U_\infty}{\mathrm{d}x}\frac{\partial}{\partial x}\left(\frac{\tau_{yx}}{\rho U_\infty^2}\right) + \frac{\partial^2}{\partial x^2}\left(\frac{\tau_{yx}}{\rho U_\infty^2}\right)\right]\mathrm{d}y\mathrm{d}y,
\end{aligned} \tag{A.14}$$

$$B = \frac{4}{U_\infty}\frac{\mathrm{d}U_\infty}{\mathrm{d}x}\int_0^{\delta_t}\int_{\delta_t}^{y}\left[\frac{\partial}{\partial x}\left(\frac{\bar{u}\bar{v}}{U_\infty^2}\right) - 2\frac{\bar{v}}{U_\infty}\frac{\partial}{\partial x}\left(\frac{\bar{u}}{U_\infty}\right)\right]\mathrm{d}y\mathrm{d}y$$
$$+ 2\int_0^{\delta_t}\int_{\delta_t}^{y}\left[\frac{\bar{u}}{U_\infty}\frac{\partial^2}{\partial x^2}\left(\frac{\bar{v}}{U_\infty}\right) - \frac{\bar{v}}{U_\infty}\frac{\partial^2}{\partial x^2}\left(\frac{\bar{u}}{U_\infty}\right)\right]\mathrm{d}y\mathrm{d}y, \quad \text{(A.15)}$$

are applied.

A.2 Influence of Domain Size on Drag Reduction Rate

This section presents the investigation of the simulation domain size influence on the resultant control efficiency. The opposition control scheme for the wall-normal velocity component (equation 2.64) is applied to the entire wall area of the controlled simulations ($d = 1$) with a sensing plane position at $y_s^+ = 10$ and amplitude amplification $\alpha = 1.0$. Configuration 4 from Table 3.1 at $\mathrm{Re}_\tau = 200$ is considered as a base case for the parametric study. An investigation of the domain size variation separately in the x- and z-directions, as well as a simultaneous change to both of the domain extents is considered. The grid resolution is kept constant for all simulations, while the time period of statistical integration is increased for smaller domains in order to be able to maintain the same amount of statistically averaged samples. Hence, the ratio of statistical integration time of the reduced size cases to the integration time of the base case is inversely proportional to the domain scaling factor:

$$\frac{\Delta t}{\Delta t_b} = \begin{cases} \frac{L_{x,b}}{L_x} & \text{for streamwise reduction,} \\ \frac{L_{z,b}}{L_z} & \text{for spanwise reduction,} \\ \frac{L_{x,b}L_{z,b}}{L_x L_z} & \text{for streamwise and spanwise reduction,} \end{cases} \quad \text{(A.16)}$$

with subscript of "b" representing the base case with $(L_x^+ \times L_y^+ \times L_z^+) = (1571{\times}400{\times}628)$ and $\Delta t_b^+ = 30000$ corresponding to about 75 eddy turnovers.

Figure A.1 shows the change in the estimated drag reduction rate for various reduced domain sizes. The distributions shown also include the uncertainty estimation proposed by Oliver *et al.* [100] based on the auto regressive time series model [145]. A reduction of the domain size in the streamwise direction leads to an overestimation of the resulting R, while the reduction of spanwise extent exhibits an opposite effect. The deterioration becomes significant when the domain becomes lower than the size of $L_x^+ = 600$ or $L_z^+ = 250$. Considering a variation of the simulation domain in one direction, the smallest cases are $(L_x^+ \times L_y^+ \times L_z^+) = (198 \times 400 \times 628)$ and $(L_x^+ \times L_y^+ \times L_z^+) = (1571 \times 400 \times 98)$ for size reduction in the x- and z- directions, respectively. For these sizes, overestimation of R by 4% for x-reduction and underestimation by 2% for z-reduction is observed. Further reduction leads to a relaminarisation of the flow field in controlled or uncontrolled cases and is therefore excluded from the plot. An increase in the estimated uncertainty is also evident for smaller domain sizes, especially for extremely small ones. A combined reduction of the domain extents does not show such strong deviation of the estimated drag reduction rate: the maximum discrepancy in R compared with the base case does not exceed 1.5%. The overestimation of R due to the reduction of streamwise domain extent is apparently cancelled by the underestimation of R due to reduced spanwise domain extent, so the deterioration of R becomes smaller. The smallest feasible case for combined reduction size is $(L_x^+ \times L_y^+ \times L_z^+) = (294 \times 400 \times 128)$ corresponding to a scaling factor of 0.2, where, however, a very high uncertainty is observed.

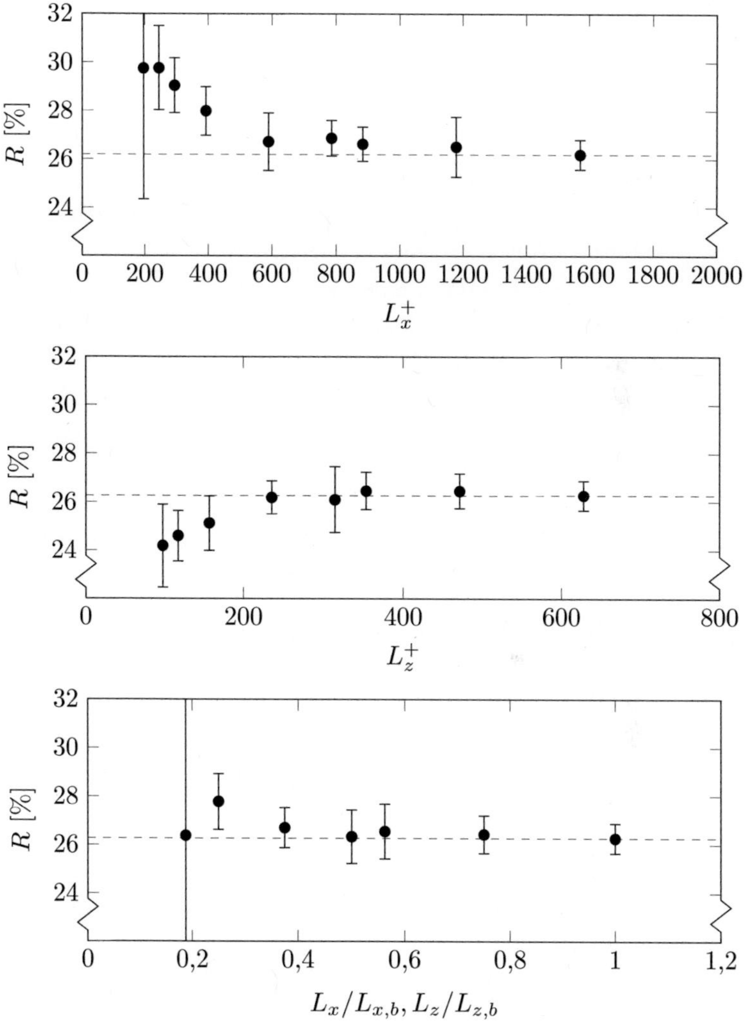

Figure A.1: Influence of the domain size on the resulting drag reduction rate. Upper and middle graphs represent reduction of x- and z-extents, respectively. Lower graph shows simultaneous reduction of both domain extents, while the scaling factor of 1 represents the base configuration 4 from Table 3.1.

A.3 Influence of Actuator Properties on Control Performance

This section provides a summary of the investigations linked to the reduction of the control scheme resolution in TCF. The reduction in the resolution is implemented utilising finite actuator dimensions and separations. The actuator size is varied from $\Delta x_a^+ = 9.8$ to 588 in the streamwise and from $\Delta z_a^+ = 3.9$ to 58.5 in the spanwise directions. Every finite size actuator with dimensions $(\Delta x_a \times \Delta z_a)$ acquires its control input from one sensor placed in the midpoint of the corresponding actuators at the sensing plane, so an increase of actuator size always means a reduction of the spatial sensor resolution. Configuration 2 (Table 3.1) at $\mathrm{Re}_\tau = 150$ is used as a base simulation set-up. The streamwise and spanwise amounts of grid nodes is set to $N_x = N_z = 120$ in order to broaden the spectrum of integer factors that can be used for the spatial definition of an actuator. The opposition control scheme for the wall-normal velocity component (equation (2.64)) is employed on the entire TCF wall area ($d = 1$) for the study with a sensing plane position at $y^+ = 10$. Simulations are carried out under CFR condition and the amplification factor is set to $\alpha = 1$. Statistical integration is performed during $\Delta t^+ = 13000$ or 40 eddy turnovers.

Figure A.2 shows the control performance for the separate increase of actuator size in the x- or z-directions while the extent in the opposite direction matches the resolution of the simulation domain.

Furthermore, a parametric study with simultaneous variation of actuator size in both directions is carried out. Figure A.3 depicts the results in terms of R, S and G. Previously presented distributions from Figure A.2 can be observed in this figure as a vertical line for $\Delta x_a^+ = 9.8$ or a horizontal line for $\Delta z_a^+ = 3.9$.

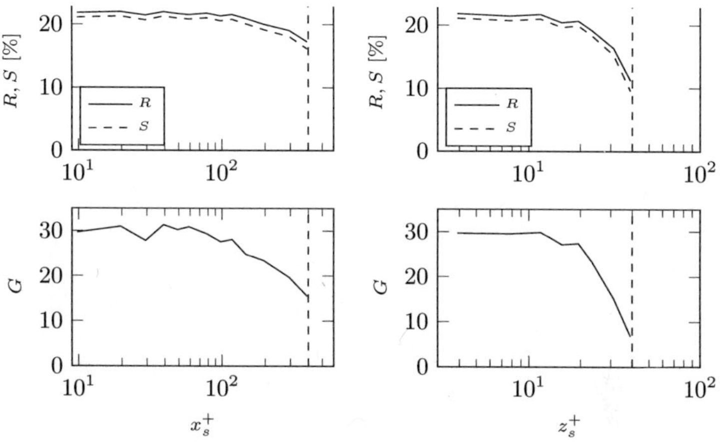

Figure A.2: Control performance of opposition control scheme for actuator size increased separately in the x- or z-directions. Variation of Δx_a^+ and Δz_a^+ is performed for constant $\Delta z_a^+ = 3.9$ and $\Delta x_a^+ = 9.8$, respectively.

In general, the combination of streamwise and spanwise increase of actuator size amplifies the deterioration of control performance. Moderate reduction of control performance ($R \approx 15$, $G \approx 20$) in comparison with the full resolution actuation is observed for $(\Delta x_a^+ \times \Delta z_a^+) < (200 \times 20)$. For the largest actuator size of $(\Delta x_a^+ \times \Delta z_a^+) = (400 \times 40)$ it is still possible to yield $R \approx 8\%$, $S \approx 6\%$ and $G \approx 6$.

As a next step, a configuration with spatial separation of actuators is considered. The separation of actuators is introduced with equidistant gaps of $\Delta x_{a,s}$ and $\Delta z_{a,s}$ in streamwise and spanwise directions. Obviously, the increase of the actuator separation leads to a reduction of the area with control input. Since the sensor configuration is kept similar to the previous configuration with a sensor associated with an underlying actuator, the sensing resolution is also reduced. The representative actuator size is chosen to be $(\Delta x_a^+ \times \Delta z_a^+) = (200 \times 20)$, which yields $R = 14.9\%$, $S = 14\%$ and $G = 18$ without separation of the actuators. A variation of streamwise separation with $\Delta x_{a,s}^+ = 0 - 200$ and spanwise separation $\Delta z_{a,s}^+ = 0 - 60$ is considered in a parametric study.

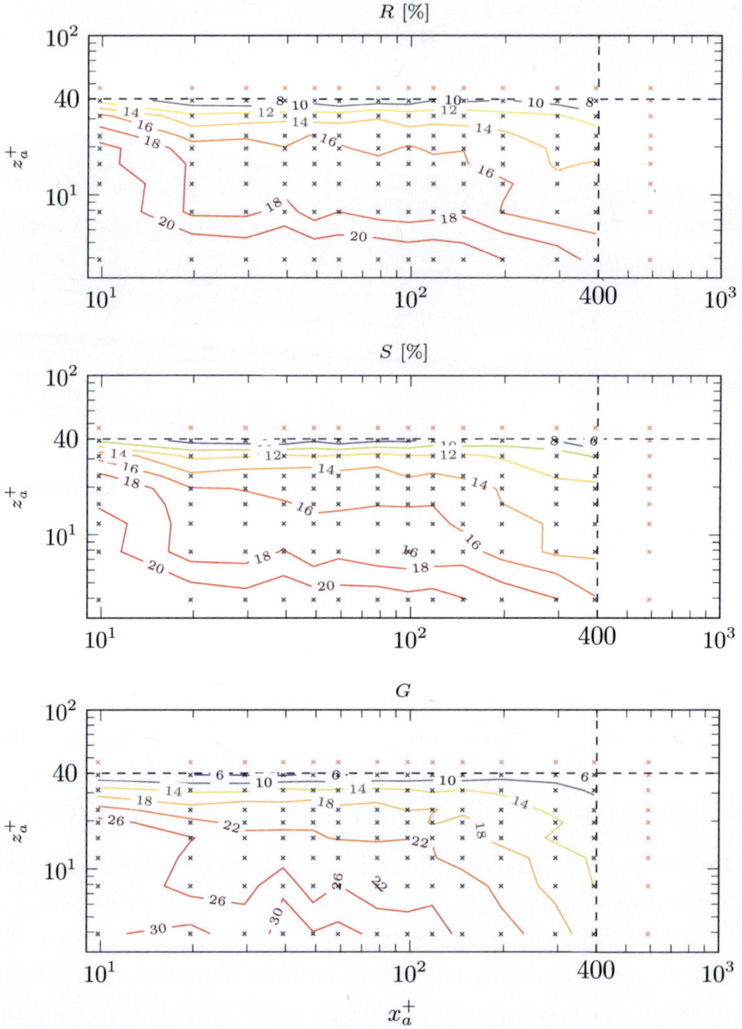

Figure A.3: Control performance of opposition control scheme for actuator size increased simultaneously in the x- and z-directions. Marks × denote a configuration carried out and × represent a failed simulation.

Figure A.4 depicts the change of the achieved control performance for increased separation $\Delta x_{a,s}$ with no separation in the z-direction and *vice versa*. In order to take into account the reduction of the active control area, a weighted drag reduction, R', is introduced:

$$R' = R_{\mathrm{fs}} \cdot \frac{A_{\mathrm{con}}}{A_{\mathrm{chan}}}, \tag{A.17}$$

where R_{fs} represents the base case with finite size actuation without separation, A_{con} is the area covered by actuators and A_{chan} is the entire wall area of the TCF. For both cases of streamwise and spanwise separation the distributions of the actual drag reduction and energy saving rate follow the distribution prescribed by the weighted drag reduction rate. However, the energy gain shows weaker deterioration, especially for the streamwise separation. The results repeatedly highlight the streamwise elongation of the affected vortical structures in the near-wall region, so the control is much more sensitive to the spanwise separation of the actuators in terms of viscous units.

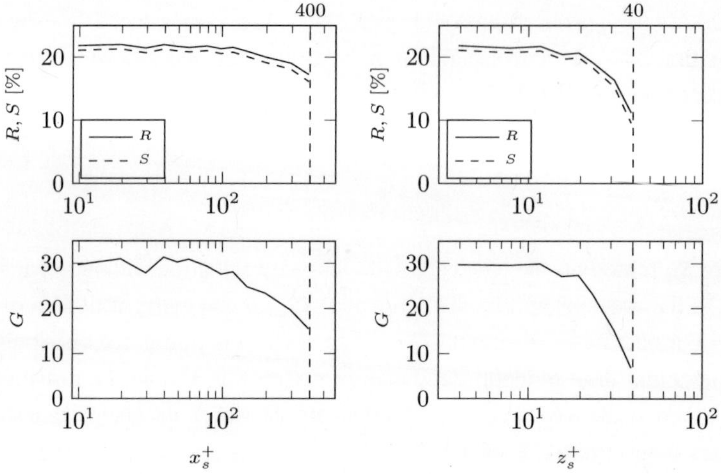

Figure A.4: Control performance of an opposition control scheme with introduced streamwise ($\Delta z_{a,s} = 0$) and spanwise separation ($\Delta x_{a,s} = 0$) of actuators with a constant size of $(\Delta x_a^+ \times \Delta z_a^+) = (200 \times 20)$. R' represents the maximal drag reduction rate without separation weighted with the control area coverage ratio.

Table A.1 shows the influence of the simultaneous extension of the streamwise and spanwise separation between single actuators. Interestingly, even for the largest separation the energy gain remains higher than 10, however the achieved drag reduction is negligible with $R < 2\%$.

$(\Delta x_{a,s}^+ \times \Delta z_{a,s}^+)$	(0×0)	(0×12)	(100×20)	(200×20)	(200×60)
control area [%]	100	65.5	50	33.3	12.5
R[%]	14.9	9.9	8.1	5.8	1.8
R'[%]	14.9	9.3	7.4	5.0	1.9
S[%]	14.0	9.2	7.5	5.4	1.6
G	18	14.6	14.2	13.7	10.3

Table A.1: Control performance for configuration with simultaneous increase of streamwise and spanwise actuator separation with fixed actuator size $(\Delta x_a^+ \times \Delta z_a^+) = (200 \times 20)$.

A.4 Influence of Sensor Distribution on Control Performance

This section elucidates the possibilities of wall sensing which can be utilized in the configuration of TCF. In order to approach the goal of wall sensing with spatially separated actuators and sensors (Figure 3.19) three intermediate configurations with sensing plane information utilising partial control, control with reduced sensing resolution and partial control with reduced sensing resolution are considered and evaluated. Eventually, in the fourth configuration the sensing plane information is replaced with wall sensing. Domain set-up 1 and 6 from Table 3.1 at $\mathrm{Re}_\tau = 150$ and $\mathrm{Re}_\tau = 300$ are utilised for the investigation of the wall sensing issue. Statistical integration is performed during $\Delta t^+ = 8500$ corresponding to a minimum of 15 eddy turnovers. The simulations are carried out under CFR conditions.

A.4.1 Partial Control

The first configuration aims at the investigation of the partial control effect. The schematic of this configuration is illustrated in Figure A.5. The wall area is divided into sections with applied control (length Δx_a) and recovery sections (length Δx_r) where the control is deactivated. The sensing is initially kept in the original form, that is, continuous information at the sensing plane, y_s, is made available. Since the length of the control area, Δx_a, is equal to the recovery area length, Δx_r, the control is always applied to 50% of the entire wall area. In the case of partial control the control input is defined as

$$\begin{aligned} v\,(x,0,z,t) &= -v\,(x,y_s,z,t) \\ \text{for}\quad \Delta x_a\cdot(2n-1) &< x \leq 2\Delta x_a\cdot n \\ \text{and}\quad n &= 1,2,3\ldots \end{aligned} \tag{A.18}$$

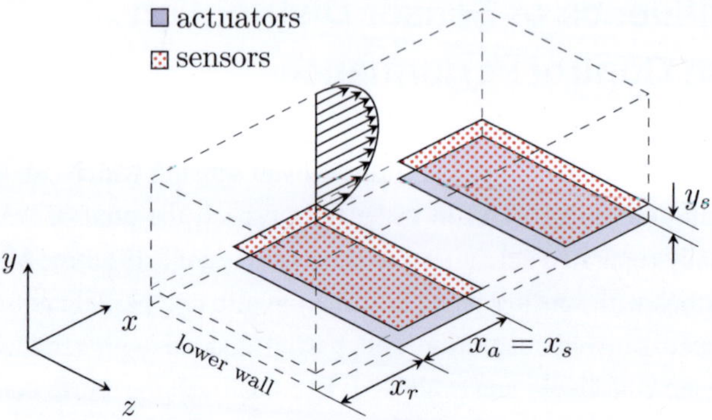

Figure A.5: Schematic of partial opposition control.

Figure A.6 summarises the control performance for the configuration where opposition control is partially applied. As a reference for the interpretation of the following result, it has to be mentioned that R = 23.5%, S = 22.8% and G = 31.9 are obtained when opposition control is applied over the whole surface. The control results show almost constant $R \approx S \approx 14\%$ and $G \approx 25$ regardless of the streamwise extent of the control and recovery areas. The achieved performance slightly exceeds 50% of the values obtained for total wall area control. The local drag (i.e., wall shear stress) gradually adjusts to the controlled or uncontrolled state as shown in Figure A.8. This transient behaviour leads to the fact that longer Δx_a exhibit lower local τ_w in the control region and higher local τ_w in the recovery region. However, the total mean value of τ_w remains almost constant for different Δx_a. This is in agreement with the results in a turbulent pipe flow with partial control as presented by Fukagata *et al.* [35].

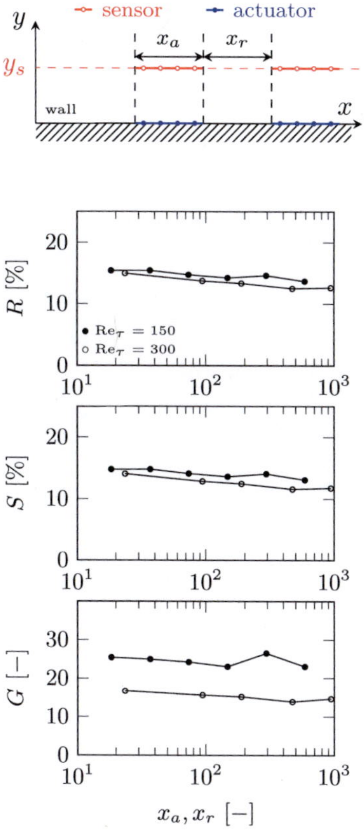

Figure A.6: Schematic of partial control configuration with control input based on sensing plane information with corresponding results for variation of control area length.

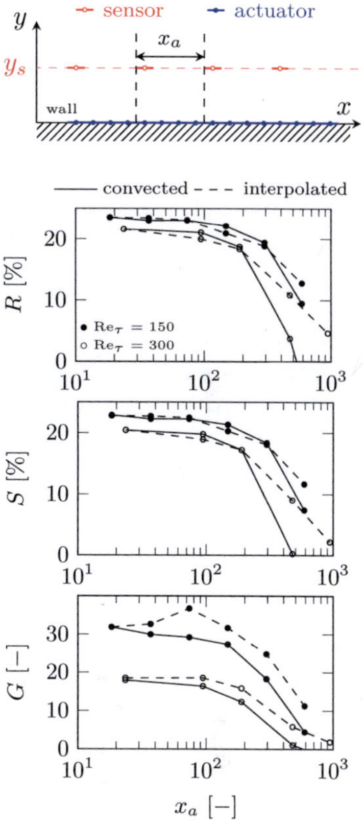

Figure A.7: Schematic of convected control configuration with control input based on local sensing plane information with corresponding results for variation of sensing resolution.

A.4.2 Reduction of Spatial Sensing Resolution

The second configuration introduces a sparser placement of the sensors along the sensing plane, y_s. In this configuration with reduced spatial sensing resolution, the total wall area is covered with actuators, as illustrated in Figure A.9. The control input is estimated by convecting the information from the sensors downstream (as described in Section 3.3) with the convection velocity $U_c^+ \approx 10$ [134]:

$$\begin{aligned} v(x, 0, z, t) &= -v(\Delta x_s \cdot (n-1), y_s, z, t - t_0) \\ t_0 &= (x - \Delta x_s \cdot (n-1))/U_c \\ \text{for} \quad & \Delta x_s \cdot (n-1) < x \leq \Delta x_s \cdot n \\ \text{and} \quad & n = 1, 2, 3 \ldots \end{aligned} \tag{A.19}$$

Another possibility of the control input estimation is a linear interpolation of the sensor information. In this case a sensor at the beginning of the control area and a sensor further downstream are used for estimation of the control input:

$$\begin{aligned} v(x, 0, z, t) &= -v(\Delta x_s \cdot (n-1), y_s, z, t) - \beta \cdot (x - \Delta x_s \cdot (n-1)) \\ \beta &= 0.5\,(v(\Delta x_s \cdot n, y_s, z, t) - v(\Delta x_s \cdot (n-1), y_s, z, t)) \\ \text{for} \quad & \Delta x_s \cdot (n-1) < x \leq \Delta x_s \cdot n \\ \text{and} \quad & n = 1, 2, 3 \ldots \end{aligned} \tag{A.20}$$

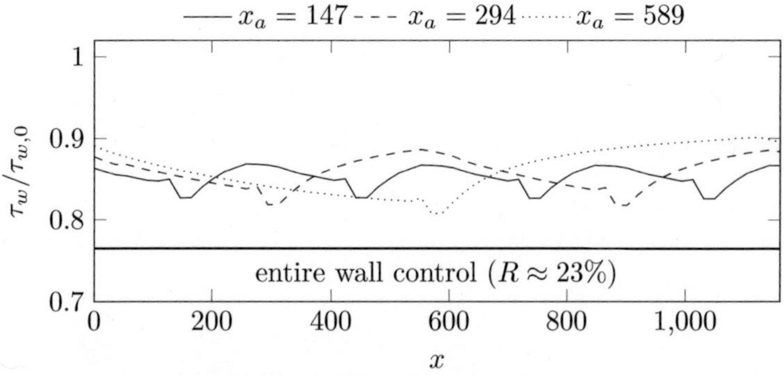

Figure A.8: Streamwise distribution of the local wall shear stress for partial opposition control with different control area lengths.

The distance between two sensors is given by Δx_s. In the following sections these estimation schemes are referred to as *convected* and *interpolated* control input, respectively.

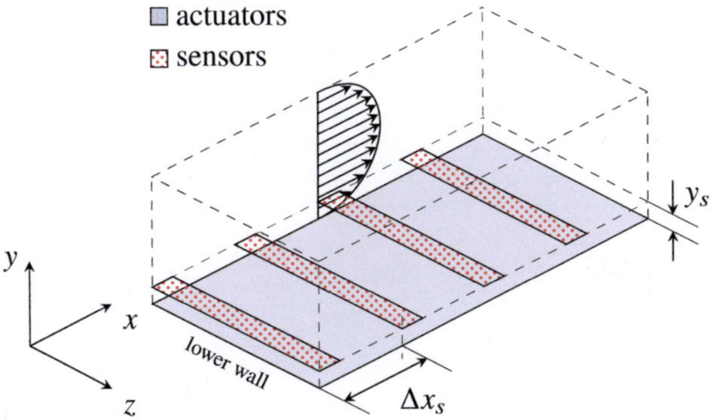

Figure A.9: Schematic of opposition control with reduced spatial sensing resolution.

Figure A.10 shows the difference in the streamwise autocorrelation for different types of the control input estimation in an uncontrolled flow $C_s\,(\phi(x, y, z), \phi(x + \Delta x, y, z))$ where ϕ represents a sensing signal processed by a certain control input estimation scheme. The convected and interpolated sensor signal for v demonstrates considerably longer integral length scales than the original plain sensor signal. The integral length scale of the signal, L_s, is computed according to equation (2.21). In viscous units $L^{+,c}_{x,v} \approx 260$ and $L^{+,i}_{x,v} \approx 100$ are found, while the integral length scale of the plain sensor signal is $L^{+}_{x,v} \approx 60$. Similar behaviour is observed for $\partial\tau_x/\partial z$, namely $L^{c}_{x,\tau}$ and $L^{i}_{x,\tau}$ are larger than $L_{x,\tau}$. It has to be mentioned that the autocorrelation for the interpolated scheme also depends on the distance between the sensors Δx_a.

Results for the configuration with reduced sensor resolution are shown in Figure A.7. Both the convected and the interpolated control inputs achieve a high drag reduction rate of $R \approx 22\%$ for a control area size of up to $\Delta x^+_a \approx 150$.

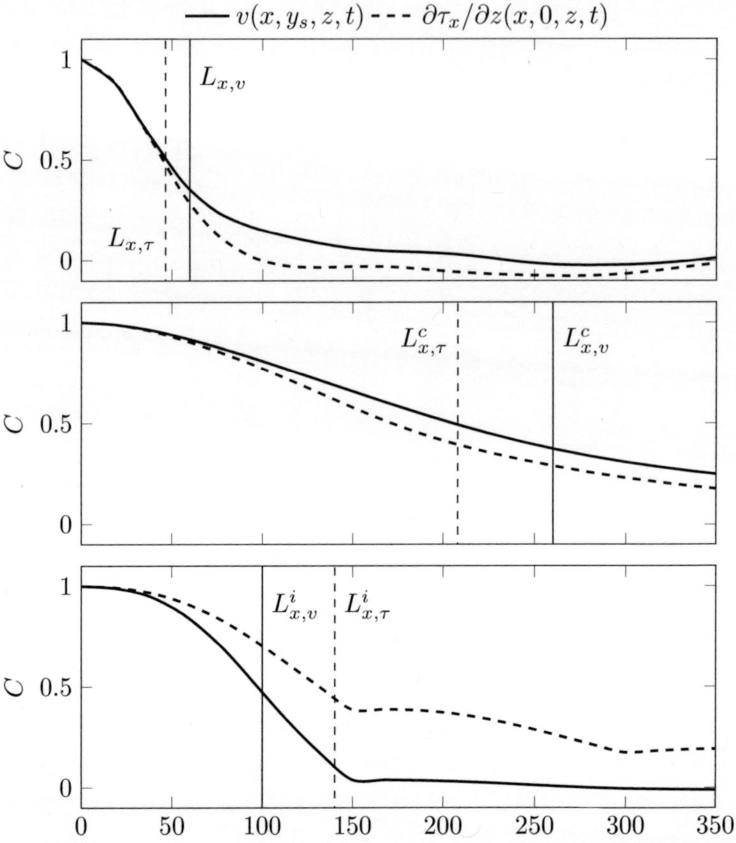

Figure A.10: Streamwise autocorrelation of plain sensor signal (upper plot), convected sensor signal ($\Delta x_s^+ = 150$, middle plot) and interpolated sensor signal ($\Delta x_s^+ = 150$, lower plot) in uncontrolled flow.

The drag reduction rate decreases for larger control areas. The decreasing performance for large Δx_a can also be observed in the energy saving rate and the energy gain. Interestingly, the interpolated control input yields the best performance in terms of S and G for a control area length of $\Delta x_a^+ = 74$. This is probably linked to the spatial filtering of sensor information, which is introduced by the interpolation procedure, leading to reduced fluctuation in the local control input and thus reduced power input. Similar behaviour is observed in previous studies where a temporal reduction of sensor resolution is applied [134].

A.4.3 Partial Control with Reduced Sensing Resolution

The third configuration constitutes a combination of the two previously introduced control schemes where partial control and reduction of the sensor resolution are simultaneously considered as shown in Figure A.11. Spatially alternating controlled, and uncontrolled regions are implemented such that half of the channel wall is covered with actuators. The sensors are spatially separated from actuators and placed at a distance Δx_u upstream. The control input is estimated using either the convected or the interpolated control input. The sensor is always placed in the uncontrolled area, once close to its end, that is, in the closest upstream position before the controlled area, and once in such a way that the furthest distance from the controlled area is realised (worst case scenario). For the first set-up the sensor is placed $\Delta x_u^+ = 18.4$ wall units upstream of the controlled area. In the second set-up we choose $\Delta x_u = \Delta x_r$ for the convected control and $\Delta x_u = 0.5\Delta x_r$ for the interpolated control. The results for different Δx_a and Δx_r are presented in Figure A.12. For the sensor placement at $\Delta x_u^+ = 18.4$ partial control with upstream sensors shows slightly better performance in terms of R for $\Delta x_a^+ < 150$ than the partial control presented in Figure A.6, where sensors are placed directly above each actuator location. This positive effect of the reduced sensing resolution may be explained by the fact that the sensor is located in an uncontrolled region where higher velocity fluctuations are sensed. Therefore, the resulting control input is stronger. This difference in available sensor information can be compared with shifting the sensing plane, y_s, further into the flow. Such a shift leads to increased values of R, as already mentioned in Section 2.4.2.

In comparison with the control scheme where control is applied along the entire wall and only sensing resolution is reduced, the scheme with partial control and reduced sensing resolution yields a faster decay of control performance with increasing Δx_a and Δx_r. This is mainly linked to the reduced validity of the control input influenced by the spatial separation of sensors from actuators and the reduction of the active control area. As expected, the "worst case" scenario reduces the possible applicable distances to $\Delta x_a^+ = 150$ and 80 for the interpolated and convected controls, respectively.

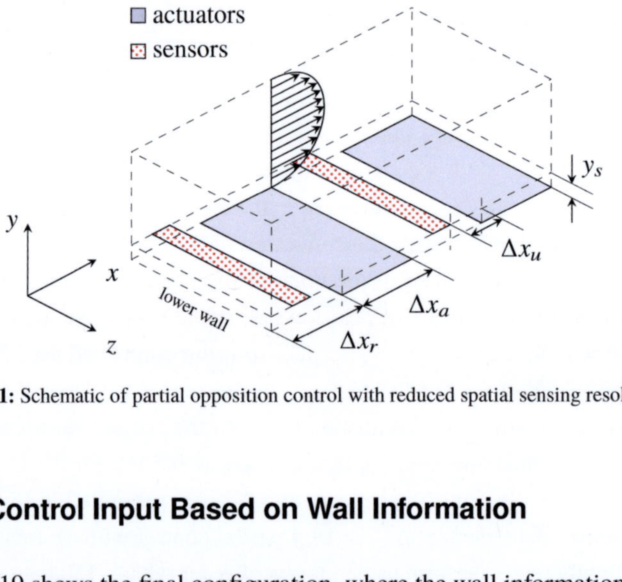

Figure A.11: Schematic of partial opposition control with reduced spatial sensing resolution.

A.4.4 Control Input Based on Wall Information

Figure 3.19 shows the final configuration, where the wall information is used to determine the control input. Wall shear stress sensors are placed within the uncontrolled area of length Δx_r, Δx_u units upstream of the control region with length Δx_a. In this configuration we also test both approaches, that is, convected and interpolated control input.

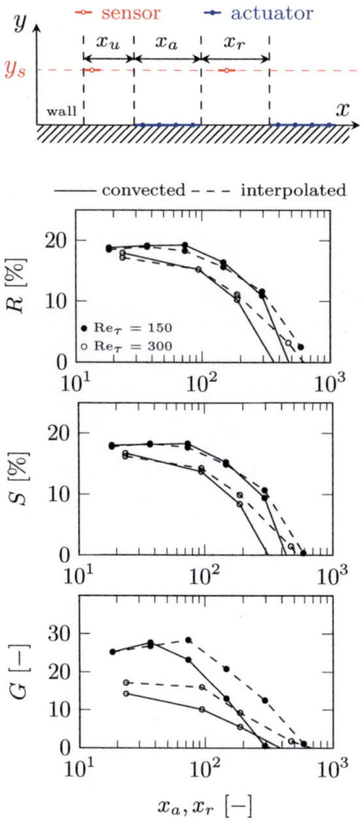

Figure A.12: Schematic of partial control configuration with control input based on local sensing plane information with corresponding results for variation of control area length and sensing resolution in terms of drag reduction and energy gain.

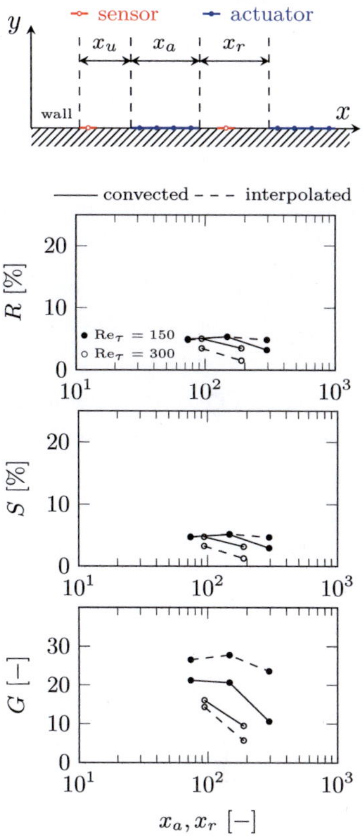

Figure A.13: Schematic of control configuration with control input based on wall information with corresponding results for optimal sensor placement and variation of control area length in terms of drag reduction and energy gain for $\alpha = 2.0$.

In order to clarify how the applied control influences the correlation between the wall shear stress and the wall-normal velocity fluctuation above the wall, the original opposition control is applied to the first half of the wall area in a streamwise direction and the two-point-correlation between v at the sensing plane and $\partial\tau_x/\partial z$ is calculated. The results are shown in Figure A.14. It is evident that the correlation recovers within the region of $\Delta x^+ \approx 74$. Due to this fact, the usage of wall information becomes possible when sensors are placed within the recovery region with a $\Delta x^+ > 74$ downstream distance from a controlled area. Therefore, we set the minimum length of the recovery region to $\Delta x_r^+ = 74$. It is found that $\Delta x_r^+ = \Delta x_a^+ = 592$ cannot achieve reasonable control performance even for the configuration with local senors above the wall (Figure A.12) due to the inability of the scheme to reproduce a valid control input for such sparse sensor distributions and long actuation areas. Therefore, three cases with streamwise spacings of $\Delta x_r^+ = \Delta x_a^+ = 74, 148$ and 296 are considered for the control scheme with wall information. It is found that the correlation completely recovers in the uncontrolled area with $\Delta x_r^+ = 148$ and 296 to $C \approx 0.4 - 0.5$ for upstream sensor positions of $\Delta x_u^+ = 54 - 74$. The shortest chosen recovery region of $\Delta x_r^+ = 74$ provides correlation values up to $C \approx 0.3$ for the upstream sensor placement with $\Delta x_u^+ = 55$.

Based on these results, a parametric study with variation of upstream sensor position Δx_u for streamwise spacing of $\Delta x_r^+ = x_a^+ = 74, 148$ and 296 is carried out. The amplification factor is set to $\alpha = 2.0$, which is found to be optimal for all three spacings. The highest values obtained for R and S are shown in Figure A.13. Table A.2 summarises the value of Δx_u at which the best control result is obtained. In Figure A.13 it is shown that the drag reduction is significantly decreased to $R \approx 5\%$. The interpolated control scheme outperforms the convected scheme in terms of G by $20 - 40\%$. In general, the upstream position of the sensor Δx_u influences the control performance in two ways. On one hand the distance between the sensors and the control section influences the estimation accuracy, which can be evaluated by the integral length scale of the sensor signal. On the other hand, Δx_u governs the correlation between $\partial\tau_x/\partial z$ and v at the sensing plane.

It is found that the sensor should be placed in an upstream position as close as possible to the control region due to the spatial decay of the autocorrelation of the sensor signal, and about 36 wall units upstream of the control region in order to capture wall information which is not influenced by the actuation. In addition, the streamwise extension of the control region, Δx_a, and the recovery section length, Δx_r, (which are set to be equally long in the present configurations) should not exceed 300 – 400 viscous units. The present control input estimation cannot provide valid flow state information for longer Δx_a and Δx_r, which is presumably linked to the integral length scale of the sensor signal.

control input	$\Delta x_a^+ = 74$	$\Delta x_a^+ = 148$	$\Delta x_a^+ = 296$
convected	36	36	36
interpolated	54	54	36

Table A.2: Best upstream position of the sensors, Δx_u, with variation of control area length, Δx_a.

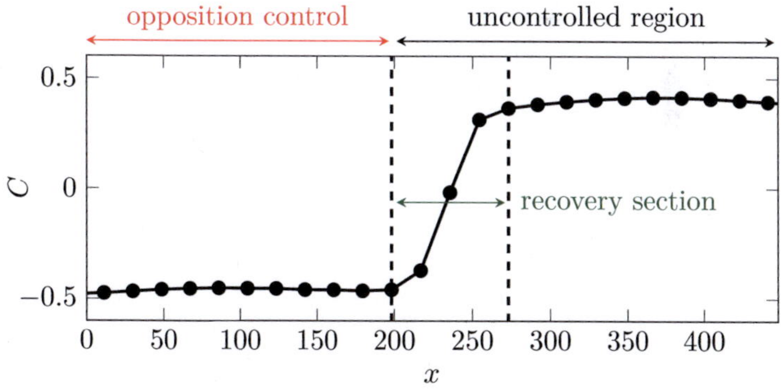

Figure A.14: Streamwise development of the correlation between wall quantity and wall-normal velocity at the sensing plane $C\,(\partial\tau_x/\partial z(x,0,z), v(x, y_s, z+\Delta z))$ for partial opposition control with half of the wall area covered by actuators and $\Delta z^+ = 15$.

A.5 Decomposition of Skin Friction Coefficient

Various types of the decomposition formulation can be derived based on the integration procedure and normalisation parameters. The conventional decomposition proposed by Fukagata, Iwamoto and Kasagi [37] utilises different velocity normalisations for different flow types, which results in an inconsistent comparison between TCF and TBL regarding decomposition of the skin friction coefficient. The following derivations are based on an internal report by Y. Hasegawa.

Fully Turbulent Channel Flow

Wall-normal integration of the non-dimensional Reynolds-averaged mean momentum equation for TCF

$$0 = -\frac{\partial \bar{p}}{\partial x} - \frac{\mathrm{d}\overline{u'v'}}{\mathrm{d}y} + \frac{1}{\mathrm{Re}} \frac{\mathrm{d}^2 \bar{u}}{\mathrm{d}y^2}, \tag{A.21}$$

yields

$$0 = -\frac{\partial \bar{p}}{\partial x} y - \overline{u'v'} + \frac{1}{\mathrm{Re}} \frac{\mathrm{d}\bar{u}}{\mathrm{d}y} - \tau_w, \tag{A.22}$$

with $\overline{u'v'} = 0$ at the wall, characteristic Reynolds number, Re, and

$$\tau_w = \frac{1}{\mathrm{Re}} \left. \frac{\mathrm{d}\bar{u}}{\mathrm{d}y} \right|_{y=0}. \tag{A.23}$$

Since pressure gradient balances the friction force in a TCF, the following relationship applies:

$$\int_0^{2\delta} \left(-\frac{\partial p}{\partial x} \right) \mathrm{d}y = -\frac{\partial p}{\partial x} = \tau_w. \tag{A.24}$$

Therefore, the relationship for the total stress is given by

$$\tau_w (1 - y) = -\overline{u'v'} + \frac{1}{\mathrm{Re}} \frac{\partial \bar{u}}{\partial y} \tag{A.25}$$

Integrating (A.22) again in the y-direction gives

$$0 = -\frac{1}{2}\frac{\partial \bar{p}}{\partial x}y^2 - \int_0^y \overline{u'v'}\mathrm{d}y + \frac{1}{\mathrm{Re}}\bar{u}(y) - \tau_w y, \tag{A.26}$$

with $\bar{u}(0) = 0$. This equation can be rewritten using (A.24) as

$$\bar{u}(y) = \mathrm{Re}\left(\tau_w\left(y - \frac{1}{2}y^2\right) + \int_0^y \overline{u'v'}\mathrm{d}y\right). \tag{A.27}$$

The centerline velocity is obtained for $y = \delta$:

$$U_c = \bar{u}(\delta) = \mathrm{Re}\left(\frac{\tau_w}{2} + \int_0^h \overline{u'v'}\mathrm{d}y\right), \tag{A.28}$$

where the first term corresponds to the centerline velocity of the laminar solution and the second one represents the loss of centerline velocity due to turbulence.

Using U_c and $\mathrm{Re}_c = U_c\delta/\nu$ for normalisation the skin friction coefficient is defined as

$$c_f = \frac{\tau_w}{0.5U_c^2}. \tag{A.29}$$

From (A.25) the following relationship is obtained:

$$\boxed{c_f = \frac{2}{\mathrm{Re}_c} - \frac{\partial p}{\partial x} - \frac{2}{\mathrm{Re}_c}\int_0^1 \overline{u'v'}\mathrm{d}y,} \tag{A.30}$$

or with (A.24):

$$\boxed{c_f = \frac{4}{\mathrm{Re}_c} - \frac{4}{\mathrm{Re}_c}\int_0^1 \overline{u'v'}\mathrm{d}y.} \tag{A.31}$$

Further integration of (A.26) from the wall to the channel center delivers

$$0 = -\frac{1}{6}\frac{\partial p}{\partial x} - \int_0^1\int_0^y \overline{u'v'}\mathrm{d}y\mathrm{d}y + \frac{1}{\mathrm{Re}}\bar{u} - \frac{\tau_w}{2}. \tag{A.32}$$

Using the mathematical relationship

$$\int_0^1\int_0^y F\mathrm{d}y\mathrm{d}y = \left(y\int_0^y F\mathrm{d}y\right)_0^1 - \int_0^1 yF\mathrm{d}y = \int_0^1 (1-y)\,F\mathrm{d}y, \tag{A.33}$$

the (A.32) can be transformed into

$$\boxed{c_f = -\frac{4}{3}\frac{\partial p}{\partial x} + \frac{8}{\mathrm{Re}_b} - 8\int_0^1 (1-y)\,\overline{u'v'}\mathrm{d}y,} \tag{A.34}$$

with the skin friction coefficient based on U_b and $\mathrm{Re}_b = 2U_b\delta/\nu$:

$$c_f = \frac{\tau_w}{0.5U_b^2}. \tag{A.35}$$

Using (A.24), equation (A.34) is expressed as

$$\boxed{c_f = \frac{12}{\mathrm{Re}_b} - 12\int_0^1 (1-y)\,\overline{u'v'}\mathrm{d}y.} \tag{A.36}$$

At the same time, if the U_c is used for normalisation, equation (A.30) can be rewritten as

$$\boxed{c_f = -\frac{4}{3}\frac{\partial p}{\partial x} + \frac{8(1-\delta_d)}{\mathrm{Re}_c} - 8\int_0^1 (1-y)\,\overline{u'v'}\mathrm{d}y,} \tag{A.37}$$

and considering substitution from equation (A.24) as

$$\boxed{c_f = \frac{6(1-\delta_d)}{\mathrm{Re}_c} - 6\int_0^1 (1-y)\,\overline{u'v'}\mathrm{d}y.} \tag{A.38}$$

Spatially Developing Turbulent Boundary Layer

Considering the Reynolds-averaged equation for the streamwise velocity component

$$\underbrace{-\left(\frac{\partial \bar{u}\bar{u}}{\partial x} + \frac{\partial \overline{u'u'}}{\partial x} + \frac{\partial \bar{p}}{\partial x} + \frac{1}{\mathrm{Re}}\frac{\partial^2 \bar{u}}{\partial x^2}\right)}_{F} = \frac{\partial \bar{u}\bar{v}}{\partial y} + \frac{\partial \overline{u'v'}}{\partial y} - \frac{1}{\mathrm{Re}}\frac{\partial^2 \bar{u}}{\partial y^2}, \tag{A.39}$$

where

$$\mathrm{Re} = \frac{U_\infty \delta_{99}}{\nu}, \tag{A.40}$$

with boundary layer thickness δ_{99} and free-stream velocity U_∞. Double integration of equation (A.39) leads to the definition

$$\tau_w = \frac{1}{\mathrm{Re}} + \int_0^1 \int_0^y F(y)\,\mathrm{d}y\mathrm{d}y - \int_0^1 \overline{u'v'}\mathrm{d}y - \int_0^1 \bar{u}\bar{v}\mathrm{d}y, \tag{A.41}$$

which can be rewritten using equation (A.33) as

$$\boxed{c_f = \frac{2}{\mathrm{Re}} + 2\int_0^1 (1-y)F(y)\mathrm{d}y - 2\int_0^1 \overline{u'v'}\mathrm{d}y - \int_0^1 \bar{u}\bar{v}\mathrm{d}y,} \tag{A.42}$$

with

$$c_f = \frac{\tau_w}{0.5U_\infty^2}. \tag{A.43}$$

The definition is consistent with skin friction decomposition in TCF from equation (A.30). Triple integration of equation (A.39) with the same normalisation yields

$$\boxed{c_f = 2\int_0^1 (1-y)^2 F\mathrm{d}y + \frac{4(1-\delta_d)}{\mathrm{Re}} - 4\int_0^1 (1-y)\left(\overline{u'v'} + \bar{u}\bar{v}\right)\mathrm{d}y.} \tag{A.44}$$

This is the well known equation proposed by Fukagata, Iwamoto and Kasagi [37], which corresponds to equation (A.37) in TCF. Considering normalisation using U_b, the skin friction can be decomposed into

$$\boxed{c_f = 2\int_0^1 (1-y)^2 F\mathrm{d}y + \frac{4}{\mathrm{Re}_b} - 4\int_0^1 (1-y)\left(\overline{u'v'} + \bar{u}\bar{v}\right)\mathrm{d}y,} \tag{A.45}$$

with

$$c_f = \frac{\tau_w}{0.5U_b^2}, \tag{A.46}$$

which corresponds to equation A.34.